Fundamentals of Reaction Flowmeters

Fundamentals of Reaction Flowmeters

Horia Mihai Moţit

CRC Press
Taylor & Francis Group
Boca Raton London New York

CRC Press is an imprint of the
Taylor & Francis Group, an **informa** business

First edition published 2022
by CRC Press
6000 Broken Sound Parkway NW, Suite 300, Boca Raton, FL 33487-2742

and by CRC Press
2 Park Square, Milton Park, Abingdon, Oxon, OX14 4RN

CRC Press is an imprint of Taylor & Francis Group, LLC.

Library of Congress Cataloging-in-Publication Data
A catalog record has been requested for this book

ISBN: 978-1-032-02168-3 (hbk)
ISBN: 978-1-032-02169-0 (pbk)
ISBN: 978-1-003-18219-1 (ebk)

Typeset in Palatino
by MPS Limited, Dehradun

To my beloved wife, Ina, for her total support, especially during this pandemic, and in memory of my late parents.

Contents

Part III Extended Reaction Flowmeters

Part IV Conclusions

Preface

Flow measurement plays a particularly important role, accentuated lately, in the optimal management of both fast-growing industrial activities and, in general, of resources (water, petroleum products, natural gas, heat, etc.), an acute necessity due to their accelerated depletion, which requires a rigorous quantitative characterization of the fluid flowing through their transport lines.

This context of evolving worldwide needs requires the accelerate improvement and completion of the flowmeters offer, with new basic flowmeters.

Starting from this requirement, and after the analysis of functional logic of all basic types of flowmeters achieved so far, the author identified the Unitary Analytical and Structural Bases of All Flowmeters, presented in his book "Unitary Analysis, Synthesis and Classification of Flowmeters" CRC Press – Taylor & Francis Group (2018).

Then, furthermore, starting from the knowledge of these unitary bases, the author developed the "Computer-Implemented Method for Unitary Synthesis and Design of Flowmeters and of Compound Gauging Structures" (European Patent EN 3364159, granted on 2020).

This book presents the results obtained by the author at the first practical implementation of this innovative method, achieved in the last two years.

These results consist of the elaboration of the "Flow Measurment Method Based on Reaction Force", and respectively of the new types of flowmeters, named by the author "reaction flowmeters", configured according to its provisions, in different basic types, systematically correlated with each other.

Initially, the book presents the principle of this new flow measurement method, and the specific basic configuration that ensures its practical application, accordingly named "the reaction measurement system", that specifically needs to be used at the configuration of all types of reaction flowmeters.

This reaction measurement system, mandatory used by all reaction flowmeters, by its configuration, ensures both the generation of the reaction force F_R (which, according to its functional equation, is proportional to the square of the measured mass flow Q_m), and its highlighting for taking over and measuring.

Starting from these developments detailed in Part I, further on are presented the basic types of reaction flowmeters, configured by customizing application of the Computer-Implemented Method for Unitary Synthesis and Design of Flowmeters and Compound Gauging Structures, corresponding to provisions of the "Flow Measurement Method Based on Reaction Force".

Thus, initially Part II presents the results obtained in the first stage, targeting the configurations design of the individual "reaction flowmeters", and then in Part III, in the second stage of this approach, the configurations design of the special interconnections of two reaction flowmeters, named "extended reaction flowmeters".

Part II successively describes the individual reaction flowmeters without moving parts and with moving parts, being thus underlined the great advantage of the method and of its corresponding"reaction measurement system", to be universally used for both these structural groups of flowmeters.

All types of the reaction flowmeters without moving parts have the same configuration of the "reaction measurement system", but each of them has a specific mode to ensure the measurement of the reaction force F_R, by measuring its different effects, which represents the characteristic parameters (the measured parameter) of the reaction flowmeters.

Accordingly were configured three variants of the reaction flowmeters with nonlinear dependence "flow rate- measured characteristic parameter", corresponding to the specific measured effect of the reaction force, respectively: with the reaction moment measurement, with differential measurement of pushing force, and with direct measurement of pushing force.

For each of the above variants of the reaction flowmeters, the specific flow rate equations are determined, and the constructive configuration and functioning are detailed.

The analysis continues with a detailed presentation, for the first above indicated type of reaction flowmeters, of the experimental results of its prototypes testing, and their main features.

Experimental calibration of these prototypes by means of water and air, on the one hand, found a very good matching between the experimental and the theoretical calibration, respectively a difference of up to 1% for water and up to 2% for air, and, on the other hand, ensures arguments for preliminary theoretical and experimental considerations on potential analytical conversion of flow scales of the reaction flowmeters from one fluid to another.

The configuration flexibility of the reaction measurement system ensures the possibility to use, as the characteristic parameter, the differential pressure that is proportional to the reaction force, thus being realized the second variant of this group of the reaction flowmeters without moving parts, named the reaction flowmeters with differential measurement of pushing force.

The third variant of this group, the reaction flowmeters with direct measurement of the pushing force, is configured in two subvariants of force measurement: by a load cell with strain gauge, respectively by its electromagnetic balancing.

For each of these flowmeters are similarly detailed the configuration and operation mode as well as the main features, such as the very high measurement accuracy ensured by the second subvariant.

Another variant of the group of flowmeters with reaction without moving parts is constituted by those that ensure the linear dependence "flow-measured characteristic parameter (electric voltage)", linearization obtained by ensuring the constructive functionality of the quadratic dependence between the reaction force (unmeasured parameter) and the measured characteristic parameter.

Universality and flexibility of the "reaction measurement system" is demonstrated by its remarkable facility of being realized for the reaction flowmeters without moving parts, in the same basic configuration as for reaction flowmeters with moving parts that constitutes the last presented basic type from the group of the individual reaction flowmeters.

This basic type of the reaction flowmeters, although they use the same configuration of "the reaction measurement system" as the reaction flowmeters without moving parts, have a totally different analytical characterization of the dependence "flow rate-measured characteristic parameter (the rotation frequency f of the moving part)", being accordingly determined by its functional equations and the flow factor equation.

It is continued with the presentation of their constructive configuration and operation, and then with the rendering of the results of their experimental calibration with water and air.

The bypass type reaction flowmeters end the presentation in Part II, being a combination of an individual nonlinear reaction flowmeter with a fluidic resistance placed in bypass with it.

Started from these achievements, Part III presents "extended reaction flowmeters" (abbreviated ERF), developed on the elaboration basis of the new method, named the "Extended Reaction Force Method of Flow Measurement", to ensure the maximizing of the turndown values, correlated with an optimal range of the accuracy values.

The elaborated method provides a special interconnection configuration of two reaction flowmeters without moving parts, which ensures a particular correlation between the values of their ranges, that have values of the one continuing on the other, with a very small common overlap zone between them, being also established by the algorithm with the correlated operation between the two reaction flowmeters of the extended reaction flowmeters.

It is continued with the presentation of the "extended reaction flowmeters", configured according to this method, being described their basic structural configuration and operation, respectively the determination of their functional equations (referring to characteristic pushing force, respectively to turndown).

Then are presented the specific constructive configurations achieved of the "ERF with direct measurement of characteristic pushing force" and the "ERF with the differential measurement of characteristic pushing force".

This presentation concludes with a comparative analysis between the binomial (turndown-accuracy) values of ERF and of their component reaction flowmeters, depending on their reference parameters and those of their component reaction flowmeters.

Essential advantage of these extended reaction flowmeters is the remarkable increase of the turndown values, correlated with an optimal range of the accuracy values of 0,1…2% o.r.

Finally, Part IV provides an overview on reaction flowmeter features.

In conclusion, the innovative contributions brought by the book were obtained by applying the integrated vision regarding the unitary bases of flowmeters, and the systematic way provided for practically approaching the synthesis of new types of flowmeters.

By elaborating on this new method and the new basic types of flowmeters configured according to its provisions, this pioneering work demonstrates the unitary character of flowmeters and enables the author to explore the opinions and observations of those who work in this important field of technology.

The book is written for all specialists in the field of flow measurement and instrumentation, and especially to flowmeter manufacturers and R&D specialists, in addition to the teaching staff and students at such specialized, technical, and high-level universities.

Horia Mihai Moțit

Acknowledgment

This book synthetically presents first results of a recent application of the third stage of the new integrated concept on flowmeters, elaborated by the author, that in its first stage identified the Unitary Analytical and Structural Bases of All Flowmeters, and then by their using, in the second stage, he developed the Computer–Implemented Method of Unitary Synthesis and Design of Flowmeters and of Compound Gauging Structures.

Being the result of these successive logical stages, the drafting of the book is largely due to the "idea-exchanging meetings" held by author with reputable specialists in the field of the flow measurement, within his R&D activity.

Therefore, I consider that it is a moral obligation for me to express, in this way, gratitude for their support.

Due to the essential logical connection between these stages, initially I express my thanks to reputable specialists and members of the IMEKO –TC9 (Flow Measurement Committee), who by their comments to my previous synthetic material, regarding my new concept on flowmeters, encouraged me to publish the book "Unitary Analysis, Synthesis and Classification of Flow Meters" (2018).

In this respect, I pay homage to the memory of Dr. Jean Pierre Vallet (former chief executive officer of CESAME EXADEBIT SA- France), and I am grateful to Prof. Yan zou Sun (China, former member of the Editorial and Advisory Committee of Flow Measurement and Instrumentation Journal) and to Dr. John Wright (National Institute of Standards and Technology – U.S.A.).

Also, I thank to Mike Tousin (Head of market R&D Endress + Hauser Flowtec) for his constructive comments.

I express my thanks to Prof. Alvaro Silva Ribeiro (Chair of the International Flow Measurement Conference FLOMEKO2019-Lisbon, and member of the IMEKO-TC9) for his honorable decision to select for publication my paper, successfully presented at FLOMEKO2019, about the new method of flow measurement using the reaction force and the first types of reaction flowmeters achieved according to it.

I am grateful, for his constructive comments, regarding my paper, during the FLOMEKO2019 and then regarding to the draft of this book, to Dr. Pier Giorgio Spazzini (President of IMEKO -TC9).

Also, my thanks to Ernie Hauser (Chief Sales Officer of Western Energy Support & Technology – USA), and Marc Buttler (Director Application Innovation – Emerson Automation Solutions – USA), for their encouraging comments.

Finally, I am grateful to David W. Spitzer, a well-known specialist in the world and author of important books on flow measurement, for his constructive and useful comments regarding the draft of this book.

Also, I would like to thank very much Nora Konopka, Editorial Director – Engineering Taylor & Francis Group/CRC Press, for her kind proposal to write this book focused in this new development in the flow measurement field, and for her support with high professionalism, respectively Glenon C. Butler Jr., Production Editor, for his overseing of my book production.

Accordingly, I consider that all these above mentioned persons are somehow the "moral authors" of this book.

To all of them, and to my close collaborators in the arrangement of this book from Romania, I am addressing all my thanks. Last but not least, I would like to express my full appreciation to my dear wife, Ina, who understood me during my whole time in the field of flow measurement and especially while writing this book. I found in her a true supporter, during long and difficult moments of exhausting activity.

Horia Mihai Moțit

Author

Horia Mihai Moţit, PhD, graduated from the Faculty of Automatic Control and Computers – POLITECHNICA University of Bucharest and earned a PhD in Flow Measurement.

He developed a prodigious activity within over 30 years of research-design-production of a great variety of types of mature flowmeters (variable area flowmeters, insertion flowmeters, oscillating piston meters, bypass flowmeters, flow measurement structures using flumes, test stands for volumetric calibration/respectively for gravimetric calibration of flowmeters) and recently of the new imagined reaction flowmeters.

He received distinguished awards for creativity: the National Prize for Patented Inventions (2nd place) in 1985, the National Prize for Patented Inventions (1st place) in 1987. He has designed a significant number of patented inventions in the field of industrial flowmeters (then produced them in large numbers), including the world premiere of the "method for the determination of convertion curves of flow scales of gas variable area flowmeters, by using only water".

By identifying, for the first time in the world, the "unitary bases of flowmeters", he ensured for flow measurement the bases to become an independent branch of the measurement technique, presented in his book "Unitary Analysis, Synthesis and Classification of Flow Meters" (CRC Press – Taylor & Francis Group, 2018).

Dr. Moţit, by using the provisions of these unitary bases, has achieved an European Patent (2020) for his "Computer–Implemented Method for Unitary Synthesis and Design of Flowmeters and of Compound Gauging Systems".

Recently, he imagined both a new method of flow measurement, named "The Flow Measurement Method Based on Reaction Force" and successively several basic types of "Reaction Flowmeters", achieved according to it, with prototypes finalized progressively.

Dr. Moţit is also the author of several reference books in the field, of which the most relevant include the following: *Industrial Flow Measurement* (1988), *Meters* (water meters, heat meters, gas meters) (1997), and *Variable Area Flowmeters* (2006).

He developed an intense scientific activity in the field of flow measurement, by presenting more than 35 lectures to national and international symposia/congresses and publishing more than 35 articles in national technical-scientific journals.

Dr. Moţit was director of the R&D Department for flow measurement of:

- Measurement and Control Instruments Company S.A. in Otopeni (Bucharest)

- Fine Mechanics Company S.A. in Bucharest
- Mechatronics and Measuring Technique Institute in Bucharest

Also, Dr. Moțit was technical director of the Schlumberger Industries Company– Romania.

Dr. Moțit is the founding president of the Romanian Technical Standardization Committee for Flow Measurement (1990–2006); the founding president of A.A.I.R.-Control and Instrumentation Association of Romania (1990–present); and the editorial director of "*Automation and Instrumentation*" journal (1991–present).

Dr. Motit is the member and Romania's representative at IMEKO/TC9-Flow Measurement Committee (1991–present). He was the member of the Scientific Committee of the International Flow Measurement Conferences: FLOMEKO'93 and FLOMEKO 2019.

Part I

Flow Measurement Method Based on Reaction Force: Reaction Flowmeters Clasification

1

Unitary Synthesis of Flowmeters and their Unitary Design Method, Basics of Elaborating "The Flow Measurement Method Based on Reaction Force" and "The Reaction Flowmeters"

It is useful to review the previous backgrounds of successive developments that made the development of Flow Measurement Method Based on Reaction Force possible, and then the different configurations of Reaction Flowmeters achieved according to it.

In this respect, this chapter presents a short explanation of the three logically correlated stages that led to the configuration of reaction flowmeters.

At the beginning, the author observed a series of connections between different types of flowmeters regarding their basic logic to perform the flow measurement; he felt the need to investigate their causes.

He kept getting more and more interesting results, so he was led to a gradual extension of the investigation, from the initial analysis of only several types, to its completion by individual analysis of the functional logic of all basic types of flowmeters made so far.

The synthesis of the results correlation of this holistic analysis materialized in the identification of the Unitary Analytical and Structural Bases of All Flowmeters that characterize their functional logic.

The preliminary conclusions of this analysis were the subject of the document [5.1], presented at "ISO – TC30 Flow Measurement Meeting", 1998, and "appreciated as relevant", by the final meeting report and accordingly that they need to be included in the final form of the respective analyzed ISO standard.

The deepening of this analysis, including this final conclusion, was presented for the first time in the world in the author's book [1.10], which explained an integrated concept on functional logic of the flowmeters and their combinations, by synthetically rendering both their unitary analytical and structural bases, and their unitary classification bases.

Thus, all basic structural schemes of flowmeters have been established, and included in them are all types of flowmeters realized so far.

That is the best evidence and demonstration of their conceptual unity, respective of the complete correctness of the elaborated concept.

This exhaustive analysis, which also was supported with the presentation of a series of case examples, demonstrates that, for the configuration of all flowmeters made so far as well as any future type, it will be absolutely necessary to go through the same logical stages in the same sequence.

Until now, flowmeters were configured only intuitively, with implicit major negative effects (e.g., inefficient research with very long duration and large cost).

This inadequate procedure was used because no one knew the Unitary Analytical and Structural Bases of All Flowmeters and implicitly the logical stages of the configuration of flowmeters. Consequently, it was not possibile to conceive an algorithm which ensures the optimum elaboration of their configuration.

In conclusion, the identification of the Unitary Analytical and Structural Bases of All Flowmeters and the elaboration of this new concept finalized the first logical stage of this development.

Facing this situation, and starting from the conclusions indicated in the book [1.10], the author applied for an EP patent "*Computer-Implemented Method for Unitary Synthesis and Design of Flowmeters and of Compound Gauging Structures*", for which he was granted the European Patent EN 3364159 on2020 and the corresponding national validations, (e.g., U.K., Switzerland, German Patent 602018002035.7).

The algorithm of this Method ensures the configuration of flowmeters and their combination by the establishment of each of their basic structural component blocks and, finally, by connecting all these blocks, the achievement of their assembly.

This is similar to making the whole chain by connecting its links.

Thus, the second logical stage of the trajectory of this systematic and coherent development was reached.

Being provided this logistical basis, it was possible to pass to the third logical stage, rendered by the presented development, the first practical application of this method which has been achieved by author.

In this respect, he elaborated "The Flow Measurement Method Based on Reaction Force", and thus he was able to synthesize different basic types of flowmeters named "Reaction Flowmeters" and their extended configuration named "Extended Reaction Flowmeters", according to it.

After concluding these preliminary explanations, the next chapter continues with the achievements mentioned above, obtained so far in the third logical stage of this trajectory of systematic and coherent development.

2

Principle of the New Flow Measurement Method

The new elaborated method determines the flow rate measurement, based on the reaction force measurement, by providing a mandatory specific configuration of the fluid-flowing path through the flowmeters, which ensures the proper evidence and subsequent measurement of this force.

Due to this principle, the author called this method, the "Flow Measurement Method Based on Reaction Force", with the abbreviated name the Reaction Force Method, or the RF method.

The principle of this new method consists in the measurement of the fluid mass flow rate Q_m by putting in evidence the reaction force $F_R = F_R(Q_m)$ of the measured fluid, which is proportional to the square of the mass flow rate Q_m, and respectively its simultaneous measurement directly or by its different effects.

According to its principle, the new method called "Flow Measurement Method Based on Reaction Force" causes the reaction force F_R to be generated, for all types of reaction flowmeters, by using the same specific basic common configuration of the fluid-flowing path through the flowmeter, which is called *the reaction measurement system*.

This *reaction measurement system* consists of a pair of two distinct functional parts: *the inlet tube* and *the reaction element* (achieved as a reaction tube or a reaction drum).

The measuring fluid enters the flowmeter through the inlet tube, which is immobile and is fixed rigidly to the flowmeter housing, for all types of reaction flowmeters, and then is transferred, by a free coupling, to the reaction element.

The reaction element may rotate relative to a shaft that ensures its mechanical coupling with the inlet tube.

According to the momentum theorem, the force F_R exerted by the fluid on the wall of the reaction element (reaction tube or reaction drum) is proportional to the square of the measured mass flow rate Q_m.

According to this Method, the reaction element performs, by its specific configuration, two mandatory functions of the reaction flowmeters: on the one hand, it gets the maximum effect of the reaction force $F_R(Q_m)$ exerted by the measuring fluid on the reaction element, and on the other hand, it ensures the constructive facility that allows this effect of reaction force F_R to be preciously measured.

Due to its specific configuration, the reaction measurement system creates the measurement facility of the reaction force F_R, either directly or by balancing of its moment (named reaction moment M_R).

The special flexibility of this specific system is remarkable, because it can ensure the balancing of this tendency to rotate of the reaction element, both in a static way (without the displacement of the reaction element), or in a dynamic way (with the rotation of the reaction element).

Consequently, the Flow Measurement Method Based on Reaction Force ensures, by using its *reaction measurement system*, the measurement of a parameter characterizing the balancing moment of the reaction moment M_R, that is proportional to F_R value.

According to the solutions developed so far by the author and presented in this book, the measured parameter for the M_R static balancing way is force, or moment of force, or differential pressure, and respectively for the M_R dynamic balancing way, is the rotation frequency of the reaction element.

Then, finally, by knowning the analytical relationship between F_R and Q_m, the Method computes the corresponding value of the measured mass flow rate Q_m, according to the respective F_R value.

As mentioned previously, flowmeters configured using this Method are named *reaction flowmeters*.

In conclusion, this new method of flow measurement, due to its principle and the flexibility of its *reaction measurement system*, has a global application and covers by its use both configurations of the reaction flowmeters without and with moving parts.

3

Specific Reaction Measurement System of All Reaction Flowmeters

3.1 Basic Configuration of the Specific Reaction Measurement System

Figure 3.1 presents the basic configuration of *the reaction measurement system*, common to all reaction flowmeters.

The reaction measurement system consists of a pair of tubes: an immobile inlet tubes 1, configured as an extension of inlet connection of the flowmeter, and the reaction tube 2.

Following the mandatory requirements, mentioned in the previous chapter, the reaction tube 2 is specifically configured, on the one hand bent at 90° at its fluid discharge end (to solve the first requirement), and on the other hand, is provided at its inlet with a shaft 3, perpendicular on it, which ensures a potential rotation mobility of the tube around the shaft (to solve the second requirement).

So configured, the reaction tube will detect and measure the moment M_R of the reaction force F_R to its rotation shaft 3, and by its processing, the calculation of the mass flow rate Q_m.

3.2 Equation of the Reaction Force Generated by "the Reaction Measurement System"

The analysis is related to the "reference control volume" CV, marked in Figure 3.1, with faces at the inlet (section S_1 – the outlet from connection 1, with inner area A_1), and at the outlet (section S_2 – with inner area A_2 of the reaction tube 2), encompassing the reaction tube walls.

According to the momentum theorem, the rate of change of momentum through the control volume, for a fluid which has a steady flow in a

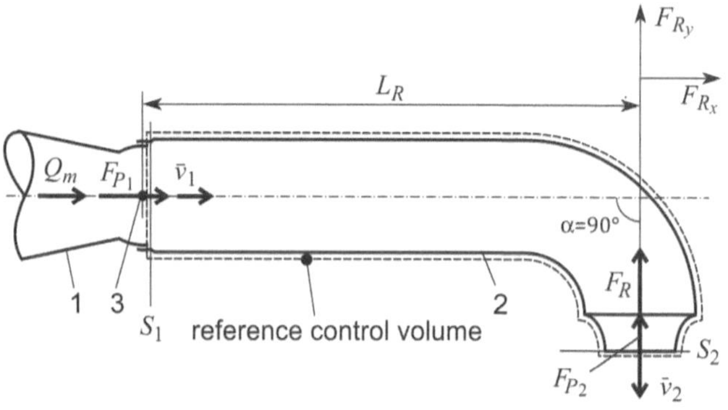

FIGURE 3.1
Configuration and the operating principle of the reaction measurement system of reaction flowmeters without moving parts.

non-uniform flowing in a stream tube, is the total force exerted on the fluid, with the vector equation:

$$Q_m \times (\bar{v}_2 - \bar{v}_1) = \bar{F}_{RT} + \bar{F}_p + \bar{F}_G \tag{3.1}$$

where:

Q_m – mass flow rate of fluid

$\bar{v}_1$ – inlet velocity of fluid into the control volume (the outlet velocity from the connection 1)

$\bar{v}_2$ – outlet velocity of fluid from the reaction tube

$\bar{F}_{RT}$ – force exerted on the fluid by the reaction tube, touching the control volume

$\bar{F}_G$ – force exerted on the fluid body (e.g., fluid weight of control volume)

$\bar{F}_p$ – force exerted on the fluid by fluid pressure outside the control volume $\bar{F}_p = \bar{F}_{p_1} + \bar{F}_{p_2}$

$\bar{F}_{p_1}$ – force exerted on the fluid at the inlet of the control volume

$\bar{F}_{p_2}$ – force exerted on the fluid at the outlet of the control volume

According to Newton's 3rd Law, regarding the "Principle of Action and Reaction", the force exerted by the fluid on the solid body (e.g., reaction tube), touching the control volume, is opposite to $\bar{F}_{RT}$ force, with the reaction force F_R given by expression $\bar{F}_R = -\bar{F}_{RT}$, and has the vector equation:

$$\bar{F}_R = Q_m \times (\bar{v}_1 - \bar{v}_2) + \bar{F}_{p_1} + \bar{F}_{p_2} + \bar{F}_G \tag{3.2}$$

Because the reaction tube shown in Figure 3.1 is placed in a horizontal plane (a two-dimensional x, y reference system), it is normal to use, instead of a vector equation, its corresponding scalar equations, respectively its components: $\bar{F}_{R_x}$ (in x-direction) and $\bar{F}_{R_y}$ (in y-direction).

It is convenient to choose the co-ordinate axis so that one is pointing in the direction of inlet velocity. So, in Figure 3.1, the x-axis points in the direction of the inlet velocity $\bar{v}_1$; as a consequence, the y-axis points in the direction of the outlet velocity $\bar{v}_2$.

On the one hand, the only fluid body force is that exerted by gravity, which acts into the paper plane, a direction that is not relevant in this analysis (for all types of reaction flowmeters that are horizontally placed).

On the other hand, it is known from the continuity equation that the volume flow rate has the equation:

$$Q_V = A_1 \times v_1 = A_2 \times v_2.$$

Generally the resultant reaction force $F_{R_{resultant}}$ is calculated by combining its components F_{R_x} and F_{R_y}.

Specifically, for the reaction tube from Figure 3.1, the component F_{R_x} of the reaction force in the x-direction does not contribute to the displacement of the reaction tube, because its effect is integrally taken over by rotation shaft 3, which ensures the mobility of the tube for its potential rotation.

Following these considerations, by using the Bernoulli equation, and by replacing v_1 and v_2 for the control volume that is open at both its ends, the resulting equation is:

$$F_{R_y} = Q_m^2 \times k_1 \times \rho^{-1} \tag{3.3}$$

where:

ρ – density of the measured fluid
$k_1 = k_1\ (A_1, A_2)$ – constructive constant.

4

Classification of Reaction Flowmeters

According to the above presentation, starting from the identification of the *Unitary Analytical and Structural Bases of All Flowmeters*, the author elaborated on the "Computer-Implemented Method of Unitary Synthesis and Design of Flowmeters and of Compound Gauging Structures" and, at its first practical implementation, he succeeded with the elaboration of a new flow measurement method (named the *FLow Measurement Method Based on Reaction Force*) and the new types of flowmeters (named *Reaction Flowmeters*), achieved in different basic structural configurations.

Before continuing with this presentation, it is necessary to reiterate the next important conclusions indicated in the book [1.10].

Global analysis of all basic types of flowmeters achieved till present identified that the vast majority have the same simple structure (called "conventional structure"), the respective flowmeters being called "conventional flowmeters", and a small minority has different structures, globally called "unconventional structures", the respective flowmeters being called "unconventional (complex) flowmeters".

Next follows a presentation of the reaction flowmeters, in a logical succession, with their various basic types, followed by the evolution of the constructive achievements of their structural blocks and the implicit consequences regarding their afferent functional parameters.

This presentation renders, at the same time, the classification of the reaction flowmeters, because both follow the same criteria of analysis of these flowmeters.

a. ***Reaction flowmeters***

According to the previous explanations, the structural scheme generally observed by all conventional flowmeters, see [1.10], is initially presented as a reference.

Consequently, conforming to these considerations and especially to the general structural scheme shown in Figure 4.1, the chapter presents the basic types of reaction flowmeters configured till present, with reference to their structures, following the progressive increase of the complexity of their configurations.

a.1 Reaction flowmeters without moving parts and nonlinear dependence "flow rate – measured characteristic parameter"

Figure 4.2 presents the structural scheme of the reaction flowmeters without moving parts and nonlinear dependence "flow rate – measured characteristic parameter".

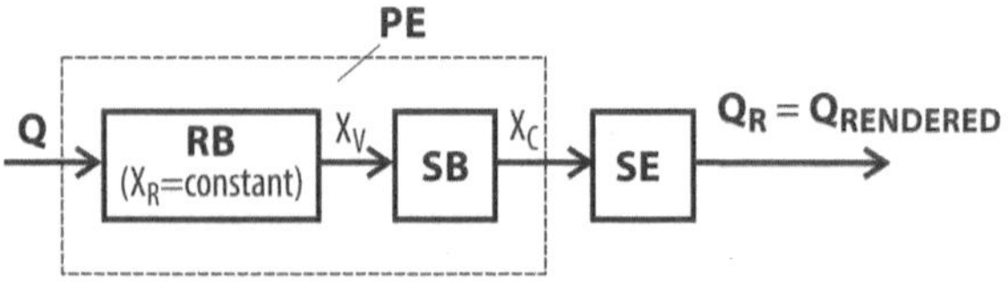

FIGURE 4.1
General structural scheme of conventional flowmeters
Legend: PE – primary element, SE – secondary element, RB – reference block (RB functionally ensures, or allows to be ensured, a constant value of the reference parameter X_R to the variation of the measured flow Q), SB – sensitive block (SB detects variation of X_V and implicitly flow rate Q variation), X_R – reference parameter (X_R is kept constant to the flow rate variation by the running mode itself of the RB), X_V – variable parameter (X_V is variable depending on the variation of flow rate Q), $X_V = Q/X_R$ – variable parameter of the flowmeter, X_C – characteristic parameter (X_C characterizes the output from PE), Q_R – parameter rendering the measured flow rate Q.

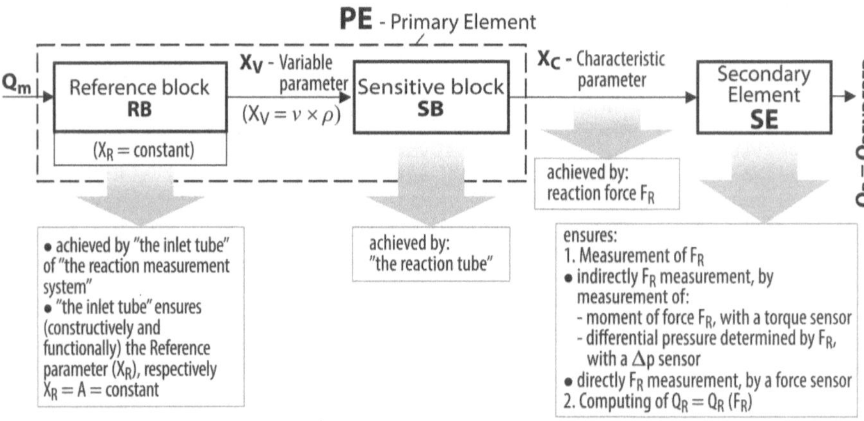

FIGURE 4.2
Structural scheme of reaction flowmeters without moving parts and nonlinear dependence "flow rate – measured characteristic parameter"
Legend: A – area of the fluid flowing through "the inlet tube", v – mean fluid velocity through "the inlet tube", respectively at inlet of "the reaction tube", ρ – fluid density, X_C – characteristic parameter (proportional to X_C, the Secondary Element SE generates the output parameter Q_R of reaction flowmeter, that renderes the measured flow rate Q_m), Q_R – parameter rendering the measured flow rate Q_m.

The Reference block RB is constructively achieved by "the inlet tube" of "the reaction measurement system", and ensures by its configuration the Reference parameter X_R and its functional role, concretely $X_R = A$ = constant.

By keeping the value of the X_R parameter constant, at the output of the Reference block RB, it generates the Variable parameter X_V proportional to the measured mass flow rate Q_m.

The Sensitive block SB receives the Variable parameter X_V and determines at its output the Characteristic parameter X_C, expressed by the reaction force F_R (generated and nonlinear depending on the mass flow rate Q_m).

The Characteristic parameter $X_C = F_R$ is in the same time the output parameter of the Primary Element PE of these basic types of the reaction flowmeters.

Then, the Secondary Element ES ensures the measurement of the reaction force F_R in three different modes, directly or by the measurement of its effects: moment of reaction force F_R, differential pressure Δp.

Accordingly, three variants of these reaction flowmeters were configured, corresponding to the specific mode to measure the reaction force F_R, respectively:

1. Indirectly, by the measurement of its effects:
   - with the reaction moment M_R measurement of the reaction force F_R, ensured by a torque sensor
   - with differential measurement of pushing force, ensured by a differential pressure sensor
2. Directly:
   - with direct measurement of pushing force, ensured by a force sensor

a.2 Reaction flowmeters without moving parts and linear dependence "flow rate – measured characteristic parameter"

Figure 4.3 presents the structural scheme of the reaction flowmeters without moving parts and linear dependence "flow rate – measured characteristic parameter".

The Reference block RB has the same configuration as the previous basic type of the reaction flowmeters presented in Figure 4.2, and the Variable parameter X_V is applied to the input of the Sensitive block SB. This structural block ensures both the generation of the reaction force F_R and its functional nonlinear conversion in the Characteristic parameter X_C.

Thus, these reaction flowmeters ensure, by the constructive configuration of Sensitive block, and not by the analytical processing, a linear dependence between the measured mass flow rate Q_m, and their characteristic parameter (respectively the electrical tension, U for the presented constructive configuration).

a.3 Reaction flowmeters with moving parts

Figure 4.4 presents the structural scheme of "reaction flowmeters with moving parts".

The Reference block RB has the same configuration as the previous basic types of the reaction flowmeters presented in Figures 4.2 and 4.3, the Variable parameter X_V being applied to the input of the Sensitive block SB.

The Sensitive block SB is made by the "rotating reaction element" consisting by the joint of two similar reaction tubes, with diametral outputs and opposite senses of the reaction forces generated.

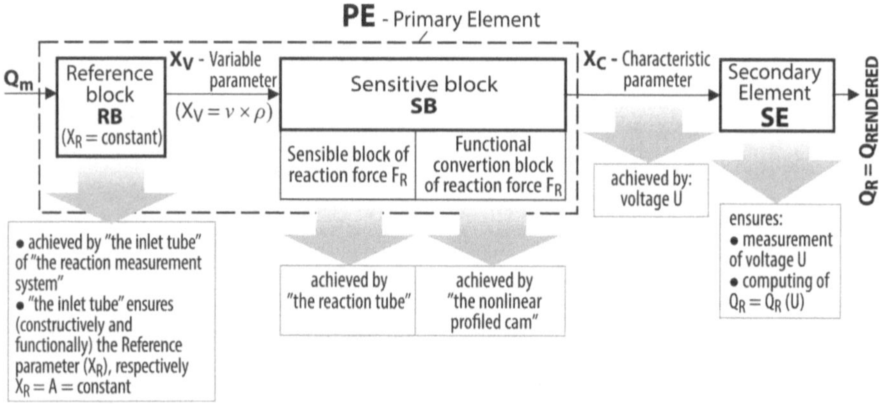

FIGURE 4.3
Structural scheme of reaction flowmeters without moving parts and linear dependence "flow rate – measured characteristic parameter".

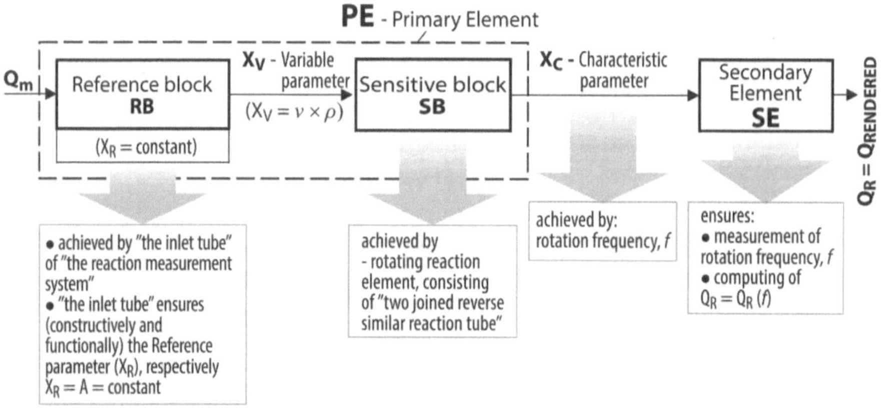

FIGURE 4.4
Structural scheme of reaction flowmeters with moving parts.

Thus, at the SB output is obtained the Characteristic parameter X_C (resectively the rotating frequency f of "the mobile reaction element").

X_C (respectively the frequency f) expresses, for each value of the mass flow rate Q_m, the achievement of the equilibrium between the two opposite torques: the driving torque of the reaction force F_R and the retarding torque τ_r of the drag.

Then, the Seconday Element ES ensures the measurement of the rotating frequency f corresponding to the measured flow rate Q_m.

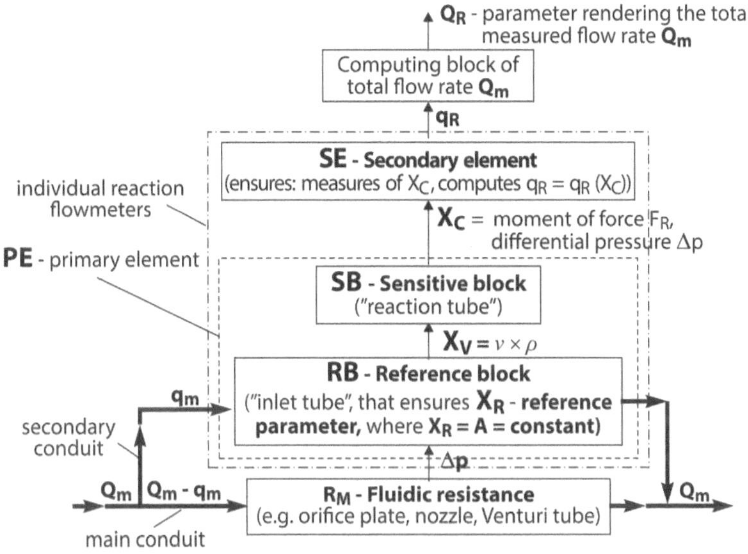

FIGURE 4.5
Structural scheme of the bypass type reaction flowmeters
Legend: Q_m – total measured flow rate, q_m – secondary flow rate, Δp – presure drop on "the fluidic resistance" (R_M) placed on main conduit, X_C – characteristic parameter, q_R – parameter rendering the measured secondary (bypass) flow rate q_m, Q_R – parameter rendering the total measured flow rate Q_m.

b. ***Bypass type reaction flowmeters***

Figure 4.5 presents the structural scheme of the *bypass type reaction flowmeters*.

The *bypass type reaction flowmeters* are achieved according to the general unconventional structural scheme, presented in the book [1.10]. This configuration ensures a closed derivation (bypass) of a secondary conduit (provided with a flowmeter that measures this small flow rate), from a main conduit on which is placed a fluidic resistance R_M.

According to this complex (unconventional) structure, it is measured only a small mass flow rate q_m that represents a part taken in closed derivation (bypass) from the total mass flow rate Q_m.

Thus, the total mass flow rate Q_m is determined by computation and corresponds to the measured flow rate q_m.

The bypass type reaction flowmeters use, for the measurement of the secondary flow rate q_m, the individual reaction flowmeters without moving parts and nonlinear dependence "flow rate – measured characteristic parameters".

According to Figure 4.5, which indicates two achievement variants of the Characteristic parameter X_C, in turn, the Secondary Element of these bypass type reaction flowmeters has two correspnding variants: with force sensor and with differential pressure sensor.

c. ***Extended reaction flowmeters***

Figure 4.6 presents the structural scheme of *the extended reaction flowmeters*.

These reaction flowmeters are configured as a special functional connection between two nonlinear reaction flowmeters having the two flow ranges correlated so that one has the values in continuation of the other, with a very small overlap between these ranges, being also established the algorithm of the correlated operation between the two component reaction flowmeters.

In this way, an important extension of the total measurement flow range of this basic type of reaction flowmeters is achieved, and consequently these flowmeters are called *extended reaction flowmeters*.

According to this algorithm, the diverter 1, commanded by the block 2, ensures the flowing switching from the main line to the secondary line and vice versa.

The unique Secondary element $SE_{M/S}$ receives, according to the operation algorithm of the extended reaction flowmeters, either the Characteristic parameter X_C^M of the "main reaction flowmeter", or the Characteristic parameter X_C^S of the "secondary reaction flowmeter".

Consequently, Secondary element $SE_{M/S}$ is functionally either "the secondary element of the main reaction flowmeter" (denoted SE_M), when the values of the measured flow rate fall within the range $Q_{S_{max}} \le Q_m \le Q_{M_{max}}$, or "the secondary element of the reaction flowmeter" (denoted SE_S), when

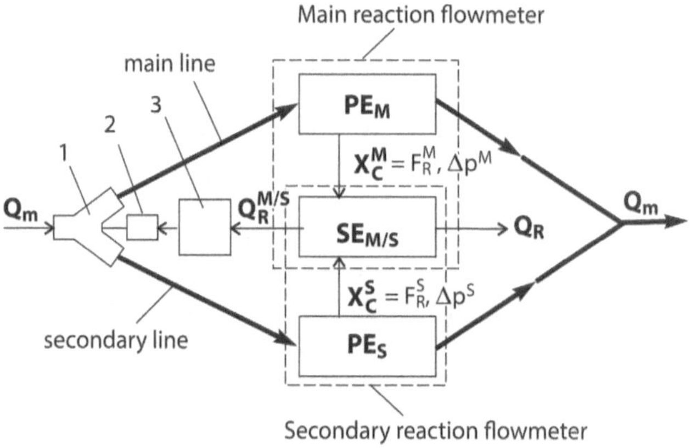

FIGURE 4.6
Structural scheme of the extended reaction flowmeters
Legend: Q_R – parameter rendering permanently the measured flow rate Q_m, PE_M – primary element of the main reaction flowmeter, PE_S – primary element of the secondary reaction flowmeter, 1 – diverter of fluid flowing from the main line to the secondary line and vice versa, 2 – actuator (solenoid) of the diverter, 3 – control block of the two positions of the diverter, X_C^M– characteristic parameter of the main reaction flowmeter, X_C^S – characteristic parameter of the secondary reaction flowmeter.

the values of measured flow rate fall within the range $Q_{S_{min}} \leq Q_m < Q_{M_{min}}$. The provision $Q_{S_{max}} > Q_{M_{min}}$ must be observed.

Similarly, $Q_R^{M/S}$ expresses the parameter that renderes, either the flow rate measured by "the main reaction flowmeter" (denoted Q_R^M), when its values fall within the range $Q_{S_{max}} \leq Q_m \leq Q_{M_{max}}$, or the flow rate measured by "the secondary reaction flowmeter" (denoted Q_R^S), when its values fall within the range $Q_{S_{min}} \leq Q_m \leq Q_{M_{min}}$.

Consequently, the algorithm of the extended reaction flowmeters ensures, by the parameter Q_R, permanent rendering of the measured flow rate Q_m (by rendering either Q_R^M or Q_R^S).

Chacarteristic parameters X_C^M and X_C^S of the main, respectively secondary, reaction flowmeter could be either the reaction force F_R^M or the differential pressure Δ_p^M.

General remark:

All types of reaction flowmeters use the same basic reaction measurement system, which is configured in two different ways, respectively:

- for reaction flowmeters *without moving parts* (the reaction element, materialized by "the reaction tube", practically immobile during the mass flow rate measurement).
- for reaction flowmeters *with moving parts* (the reaction element, materialized by "a special connection of two similar reaction tubes", mobile during the mass flow rate measurement).

Note: All types of reaction flowmeters, from each of the above group, have the coupling configuration with the process – pipe-line, correlated with the relative positions between the inlet/outlet flowmeter connections (generally with horizontal collinear axis, and only for special application, vertical collinear axis).

Later chapters continue with the presentation of all these types of the reaction flowmeters, in the sequence indicated in the above classification.

Part II

Reaction Flowmeters

5

Reaction Flowmeters Without Moving Parts and Nonlinear Dependence "Flow Rate – Measured Characteristic Parameter"

5.1 Preliminary

The measuring system of the reaction flowmeters without moving parts is achieved according to the configuration presented in Figure 3.1.

Consequently, the analysis of its operation and its functional equation refers to both Figure 3.1 and to each figure that shows the specific basic configuration variant, following the two ways of the M_R measurement presented above.

5.2 Functional Equations

A specific reaction force F_R effect corresponds to each specific basic constructive configuration of the reaction flowmeter.

In this regard, for the reaction flowmeters without moving parts, the first basic constructive configurations from the world, have been imagined and realized by the author. Flow rate equations, being specific to each basic configuration, are successively presented hereafter.

5.2.1 Flow Rate Equation of "the Reaction Flowmeters with the Reaction Moment Measurement"

This equation is deduced with reference to the Figures 3.1 and 5.1.

Moment of the reaction force F_R, which tends to rotate the reaction tube 2 about its shaft 3, which we call "the reaction moment M_R", has the equation:

$$M_R = F_{R_y} \times L_R \tag{5.1}$$

where: L_R – moment arm

The reaction moment M_R is permanently balanced by the torque τ of the torque sensor (see Figure 5.1), according to the equation:

$$M_R = \tau \tag{5.2}$$

where:

$$\tau = \theta \times C$$

C = constant (coefficient of torsional rigidity); θ – torsion angle

Essential Remark:

Because the torsion angle has an insignificant maximum value (of only $0{,}2^0 \ldots 0{,}8^0$), this type of reaction flowmeter is a flowmeter without moving parts.

By replacement of F_{R_y} in M_R expression, and by processing, it results in the following:

$$Q_m = (\rho \times \tau \times k_1^{-1} \times L_R^{-1})^{1/2} \tag{5.3}$$

5.2.2 Flow Rate Equation of "the Reaction Flowmeter with the Differential Measurement of Pushing Force"

This equation is similarly deduced as equation (5.3), but with reference to the configuration of reaction flowmeter from Figure 5.6, that ensures, proportionally to the component F_{R_y} of reaction force F_R, the force F_0 which pushes, by a pin, on a face of a separation membrane, with pushing (reaction) pressure p_0, having the expression:

$$p_0 = \frac{F_0}{A_m} = F_{R_y} \times L_R/(L_0 \times A_m) \tag{5.4}$$

where:
- L_R – arm of moment M_R, referring to reaction force F_R
- L_0 – arm of force F_0 moment
- A_m – active area of separation membrane

On the same face of the separation membrane acts, simultaneously with F_0, the static pressure p_f of measured fluid, with its F_f equivalent force.

So, the opposite face of the membrane (respectively the transmission liquid) takes over the total pressure $p_0 + p_f = (F_0 + F_f)/A_m$, and consequently the differential pressure sensor measures the difference $\Delta p = p_0$ between pressures $(p_0 + p_f)$ and p_f that acts on its (+) and (−) inlets.

By processing of the previous p_0 expression, it results in the following:

$$F_{R_y} = \Delta p \times L_0 \times A_m / L_R \tag{5.5}$$

Then, by replacing in equation (5.5) of F_{R_y} from equation (3.3) and by processing, is obtained the measured mass flow rate equation:

$$Q_m = (\rho \times \Delta p \times k_2)^{1/2} \tag{5.6}$$

where:

$k_2 = L_0 \times A_m \times L_R^{-1} \times k_1^{-1}$ – constructive constant

5.2.3 Flow Rate Equation of "the Reaction Flowmeters with Direct Measurement of the Pushing Force"

This analysis is performed with reference to Figures 3.1 and 5.7.

The moment of the reaction force F_R, which tends to rotate the horizontal reaction tube around the vertical shaft, which we name "the reaction moment M_R" has the previously indicated equation (5.1).

Given the specificity of the reaction flowmeter configuration in question, it results in the analytical relationship between the reaction force F_{R_y} and the output pushing force F_0 of the flowmeter, as follows:

$$F_{R_y} = \frac{F_0 \times L_0}{L_R} \tag{5.7}$$

where:

F_0 – output pushing force
L_R – moment arm, referring to reaction force F_R
L_0 – moment arm, referring to output pushing force F_0

Referring to Figure 5.7, the output pushing force F_0 is permanently balanced by the force sensor 26; by replacement of this specific expression of F_{R_y} in expression (3.3), it results by processing, the following flow rate equation:

$$Q_m = (\rho \times L_0 \times L_R^{-1} \times k_1^{-1})^{1/2} \times F_0^{1/2} \tag{5.8}$$

Important remark:

Because the displacement necessary, of the moving part of force F_0 sensor 26 is almost null for the measurement of the entire range ($Q_{m,min} \cdots Q_{m,max}$) of the flowmeter, correspondingly the reaction tube displacement is extremely small. *It results that this type of reaction flowmeter is practically a flowmeter without moving parts.*

5.2.4 Turndown Expression

It is known that, in flow measurement, the most commonly used term of accuracy is defined as the closeness of agreement between the measurement and the true value of the measurement.

Also, in flow rate measurement, the error is almost always usually called accuracy.

The accuracy always is expressed as the relative measuring accuracy, which is specified in two ways: as percentage of reading (% o.r.) or as percentage of full scale (% o.f.s.).

According to these considerations, the analytical correlation between "accuracy of the reaction flowmeters" and "sensor accuracy" of the reaction force F_R (measured directly or by its effects) is further presentend as being the essential characteristic of these reaction flowmeters.

This correlation is ensures with reference to the related measurement errors.

Consequently, on the one hand, considering the two ways to specify the sensor accuracy, as percentage of full scale E_S (% o.f.s.), or as percentage of reading E_S (% o.r.), results in the following:

$$E_{S,max}(\%\ o.r.) = 10^2 \times \frac{E_S(abs)}{F_{min}} = E_S(\%\ o.f.s.) \times \frac{F_{max}}{F_{min}} \tag{5.9}$$

The equation (5.9) was obtained by replacing $E_{S(abs)}$ with its expression $E_{S(abs)} = E_S(\%\ o.f.s.) \times F_{max} \times 10^{-2}$. Thus it results the equation:

$$\frac{F_{max}}{F_{min}} = E_{S,max}(\%\ \text{o.r})/E_S(\%\ \text{o.f.s.}) \tag{5.10}$$

where:

$E_{S,max}$ (% o.r.) – maximum permissible value of the relative sensor error, corresponding to the minimum measured value F_{min}

E_S (abs) – absolute sensor error, computed corresponding to relative error E_S (% o.f.s.) and the maximum measured value F_{max}

On the other hand, the value of *"the maximum permissible relative error $E_{S,max}$ (% o.r.) of the respective sensor" determines and is equal to the* *corresponding "maximum permissible relative error $E_{F,max}$ (% o.r.) of the reaction flowmeter", in whose component it is.*

In the above relations, F symbolizes the reaction force F_R, measured directly (as pushing force), or by its effects: moment of force F_R, differential pressure Δp.

Also, due to the dependence between the mass flow rate Q_m and the measured reaction force F_R, generated by it, presented by the above flow rate equations, it results for the turndown (noted "T") the following relationship for these reaction flowmeters without moving parts:

$$T = Q_{m,max}/Q_{m,min} = [E_{F,max}\,(\%\ o.r.)/E_S\,(\%\ o.f.s.)]^{1/2} \tag{5.11}$$

We need to specify that in this book, the accuracy of the reaction flow meters is expressed as a percentage of reading (% o.r.), and the sensor accuracy of the characteristic parameter, as percentage of full scale (% o.f.s.).

5.3 Reaction Flowmeters with the Reaction Moment Measurement

Related to the pipe position, these flowmeters can be achieved in two ways: with horizontal collinear inlet/outlet connections, and with perpendicular axis of inlet/outlet connections.

5.3.1 Configuration and Operation

Figure 5.1 presents the configuration of reaction flowmeter with direct measurement of the balancing torque of reaction moment and with horizontal collinear connections.

The measuring fluid enters in the flowmeter through the inlet connection 1 and continues to flow towards the reaction tube 2, bent at 90° bypass, through the spherical coupling made of the ending nozzle 3 of the inlet connection 1 and the inlet nozzle 4 of the reaction tube.

The discharged fluid from the reaction tube is taken up by a suitable convergent nozzle of the outlet connection 5 so being discharged without disturbances from the flowmeter.

Horizontal collinear connections 1 and 5 are fixed rigidly and sealed to the housing 6 of the flowmeter. Radial equidistance between the nozzles 3 and 4 is constructively achieved by two small inner bosses 7 of the nozzle 4 of the reaction tube, placed up and down around the vertical shaft 8, so being ensured an insignificant friction between the nozzles 3 and 4, having a rigorous concentric positioning accomplished by the metal shaft 8.

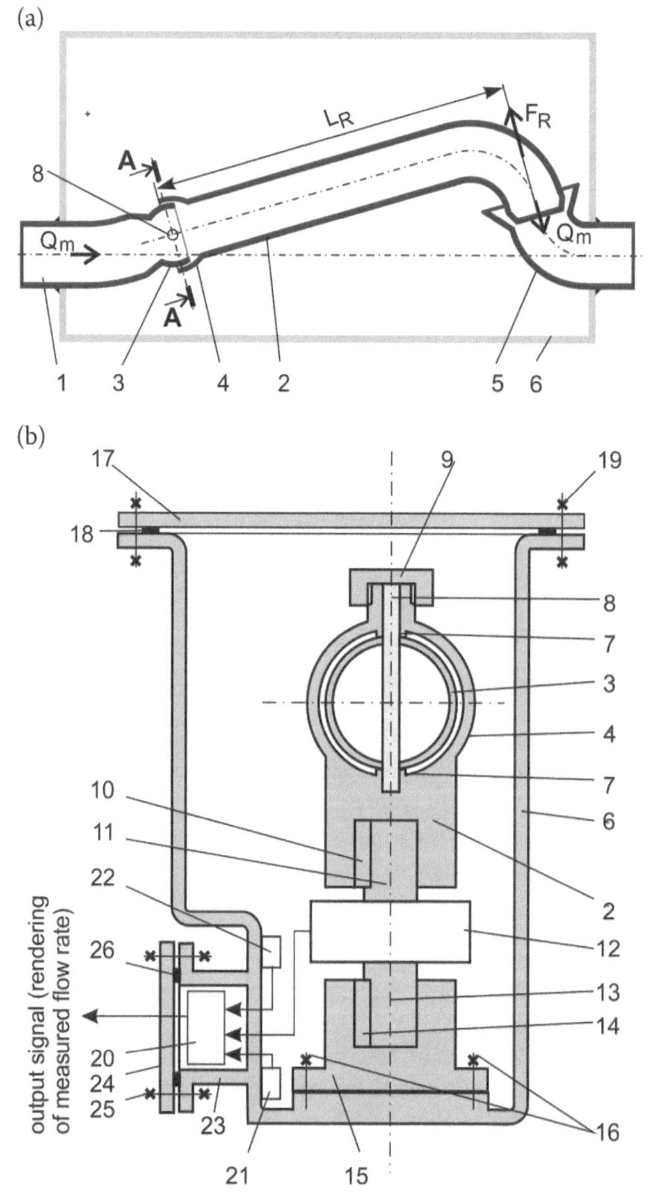

FIGURE 5.1
Reaction flowmeter with direct measurement of the reaction moment and with horizontal collinear connection. a – longitudinal section, b – cross section with plane A-A
Legend: 1 – inlet connection (immobile part of reaction measurement system), 2 – reaction tube (mobile part of reaction measurement system), 3 – ending nozzle of inlet connection, 4 – inlet nozzle of reaction tube, 5 – outlet connection, 6 – flowmeter housing, 7 – small inner boss of reaction tube, 8 – shaft of potential rotation of reaction tube, 9 – nut, 10 – wedge, 11 – measuring shaft of torque sensor, 12 – torque sensor, 13 – fixed support shaft of torque sensor, 14 – wedge, 15 – flowmeter support, 16 – screw, 17 – cover, 18 – gasket, 19 – nut/screw system, 20 – electronic block, 21 – pressure sensor, 22 – temperature sensor, 23 – outer box housing, 24 – outer box cover, 25 – nut/screw system, 26 – gasket.

Similarly, the rigorous positioning of the reaction tube with respect to the inlet connection is ensured.

By the threading of the nut 9 on the upper outer boss of the reaction tube, it is ensured a permanent locking of the shaft 8 in its rigorous position, respectively the permanent spatial positioning (x, y, z) between the nozzles 3 and 4.

By means of the wedge 10, the measuring shaft 11 of the torque sensor 12, being rigidly blocked relative to the reaction tube, respectively to its outer boss from its lower side, fully takes over the torque developed by the reaction tube.

Correct measurement of the torque is ensured by blocking the rotation of the torque transducer by locking its support shaft 13 relative to the flowmeter housing, with the wedge 14 positioned between it and the support 15, rigidly fixed to the housing by screws 16.The housing is closed and sealed by the cover 17 and gasket 18, with the nut/screw system 19.

The reaction force F_R exerted by the mass flow Q_m at the outlet of the reaction tube, determines the torque $M_R = F_R \times L_R$, proportional to the flow rate Q_m.

This rotation torque is taken up through the shaft 11 and measured by the torque sensor 12, whose output signal Q_m, proportional to Q_m, is taken over by the electronic block 20.

The shaft 11 requires a very low torsion angle of $0{,}2^0 \ldots 0{,}8^0$ to measure the maximum torque, and the reaction tube, being rigid with this shaft, has the same insignificant rotation angle for the measurement of the entire flow range ($Q_{m,min} \cdots Q_{m,max}$).

> *Thus, the rotation angle of the reaction tube is practically insignificant, and this type of reaction flowmeter is a flowmeter without moving parts.*

The electronic block renders the compensated mass flow rate Q_m with pressure P and temperature T, because in this block, on the one hand, previously was stored, the calibration curve of flowmeter $Q_m = Q_m(\tau)$ for the specific nominal operating parameters of the measured fluid, and on the other hand, it can provide the polynomial compensation of Q_m with P, and T, measured by sensors 21 and 22.

Note: We should mention that was also elaborated in a similar configuration of the reaction flowmeter, but with the indirect measurement of the balancing torque of the reaction moment from outside the measured fluid, by magnetic coupling.

5.3.2 Experimental Results. Analysis

Further there are presented the results of the experimental calibration, for these reaction flowmeters successively achieved by means of water and air.

Each experimental calibration value is presented in comparison with its corresponding theoretical calibration, which is achieved using the customized form of the equation (5.3).

Remarks:

- *In the flow measurement the most commonly used term is "the accuracy", that it is defined as "the closeness of the agreement between the measurement and the true value of that measurement". The accuracy is always the "relative measurement accuracy", as an expression of a verifiable calibration standard. There are two ways to specify "accuracy", respectively as percentage of reading of rate (% o.r.) or as percentage of full scale (% o.f.s).*
- *"The measurement error" is simply "the difference between the flowmeter output and the true value of the flow rate at the time of the measurement"; it is almost always called ~accuracy~.*
- *Consequently, it was chosen to express "the measurement precision" of all presented reaction flowmeters, throughout the entire book, by "the accuracy as percentage of reading of rate", noted E_F (% o.r.).*

5.3.2.1 Calibration with Water

5.3.2.1.1 Comparison Between Experimental and Theoretical Calibrations

With reference to Figure 5.1, a reaction flowmeter of DN25 having a torque sensor 12 (with τ_{max} = 0,5 Nm and accuracy 0,1% o.f.s.) has been calibrated with water having temperature t = (20 ± 0,2)°C by the gravimetric method. Table 5.1 presents both the values of the measured and of the calculated mass flow rate Q_m corresponding to the measured torque τ by the torque sensor 12 of the calibrated flowmeter.

Corresponding to the measured torque range (0,0102, ... 0,5) Nm of the used torque sensor, were determined the values both of the potential theoretically calculated flow rate range (310,152 ... 2171,501) kg/h, and of the corresponding experimentally measured flow rate (311,531 ... 2189,455) kg/h.

The comparison between experimental and theoretical calibration with water was concluded by the systematic calculation of their matching, expessed by the relationship:

$$\frac{Q_{m_{calc}} - Q_{m_{meas}}}{Q_{m_{meas}}} \times 10^2$$

The accuracy of this basic configuration of reaction flowmeter depends on the accuracy of the used torque sensor. Also is imposed $E_{F,\,max}$ = 2% o.r.

So, since the calibrated flowmeter used a torque sensor with τ_{max} = 0,5 Nm and accuracy of 0,1% o.f.s., it has an accuracy of (0,1% to 2% o.r.), corresponding

TABLE 5.1

Comparison between experimental and theoretical calibration with water

Torque τ Nm	Pressure drop Δp bar	Average experimental measured flow rate $Q_{m_{meas}}$ kg/h	Theoretically calculated flow rate $Q_{m_{calc}}$ kg/h	Matching $\frac{Q_{m_{calc}} - Q_{m_{meas}}}{Q_{m_{meas}}} \times 10^2$ %
0,01020	0,015	311,531	310,153	–0,44
0,02015	0,017	435,970	435,926	–0.01
0,02500	0,018	486,290	485,560	–0,15
0,03000	0,020	532,248	531,907	–0,07
0,04800	0,028	678,521	672,215	–0,93
0,07300	0,039	835,707	829,724	–0,71
0,10700	0,055	1014,740	1004,539	–0,98
0,15000	0,071	1199,002	1189,380	–0,80
0,20400	0,089	1400,101	1387,044	–0,93
0,25000	0,104	1549,122	1535,483	–0,88
0,32400	0,124	1761,315	1748,625	–0,72
0,35000	0,130	1830,906	1816,808	–0,77
0,40000	0,145	1957,716	1942,250	–0,79
0,45000	0,153	2076,891	2060,068	–0,81
0,50000	0,165	2189,455	2171,501	–0,82

to a theoretical calculated turndown of 4,47, according to equation (5.11) and relationship $Q_{m_{max}}/Q_{m_{min}} = (\tau_{max}/\tau_{min})^{1/2} = (0{,}5\ \text{Nm}/0{,}025\ \text{Nm})^{1/2} = 20^{1/2}$.

This value is confirmed and exceeded by the experimental turndown of 4,50 = 2189,455kg/h/486,290 kg/h, the ratio of the values of measured flow rate Q_m for the same values of torques 0,500 Nm, respectively 0,025 Nm. Theoretical calculation of Q_m has been achieved using the customized form of the general functional equation (5.3), according to the value of the constant k_1 determined by the construction of the calibrated reaction flowmeter, respectively:

$$Q_m = 97{,}2 \times (\rho \times \tau)^{1/2} \tag{5.12}$$

where:

Q_m – mass flow rate, in kg/h
ρ = 998,2 kg/m^3 – density of water for t = 20°C
τ – measured torque, in Nm
97,2 – customized constructive constant, in m × s

In connection with Table 5.1, Figure 5.2 plots the dependence curves of Q_m by τ, both for theoretically calculated $Q_{m_{calc}}$ and for measured $Q_{m_{meas}}$ flow

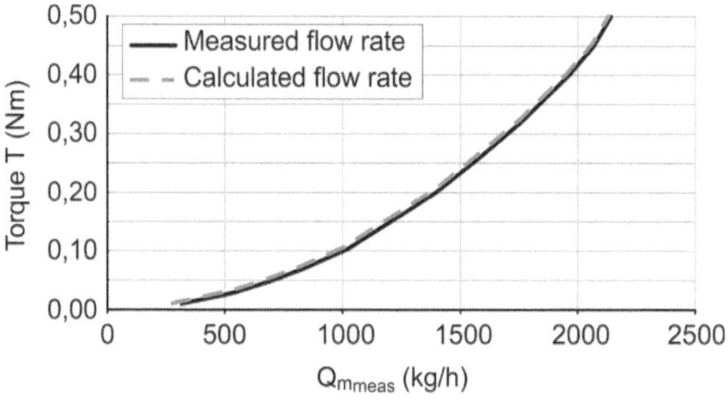

FIGURE 5.2
Mass flow rate of water depending on the measured torque (the reaction moment).

rates, the density of water being practically constant during the calibration, at t = (20 ± 0,20)°C.

5.3.2.1.2 *Main Resulting Features*

The previous analysis provides the following positive conclusions:

- Table 5.1 and Figure 5.2 demonstrate a very good matching of the experimental and theoretical calibration, the percentage differences between the values of the calculated and measured mass flow rate, being placed in a very narrow band of values of – (0,01 … 0,98)%.
- The positioning of water velocities across the flowmeter, with their maximum value (v_{max} = 1, 22 m/s), is in the center of the recommended economical velocity range.
- The pressure drop Δp depending on the flow rate has a moderate value, as presented in Figure 5.3.

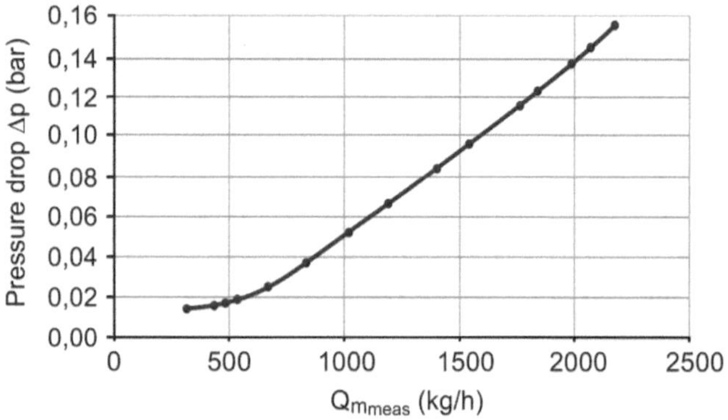

FIGURE 5.3
Pressure drop depending on the measured flow rate of water.

- Installation of these reaction flowmeters needs Inlets/Outlets of 0 × DN, which is very economical.

5.3.2.2 Calibration with Air

5.3.2.2.1 Comparison Between Experimental and Theoretical Calibrations

Another reaction flowmeter DN25 was calibrated by means of air. It has a dual-range torque sensor with measurement error 0,1% o.f.s. (that is custom-built to measure two ranges synchronously without chance-over) with the 1st range 0,5 Nm and the 2nd range 1/10 of the 1st range, respectively 0–0,05 Nm.

The master slave method was used for the calibration.

Table 5.2 presents a comparison between the average values of the measured mass flow rate (according to experimental calibration) and the theoretically calculated flow rate, with the customized equation (5.3), according to flowmeter constructive dimensions and the values of measured density ρ, corresponding to each measured torque τ, by the sensor of the calibrated flowmeter.

The rendering of this comparison, for calibration with air of reaction flowmeter, follows the same procedure used previously for its calibration with water. Consequently, in Table 5.2 is concluded this comparison by the systematic calculation of matching between the results of experimental and theoretical calibrations, using the same previous relationship as for calibration with water.

Figure 5.4 plots the dependence curves Q_m by τ, for both the theoretically calculated $Q_{m_{calc}}$, and for the experimental measured $Q_{m_{meas}}$.

Also in Table 5.2, for a complete analysis, the inlet parameters of air and the pressure drop are presented for each value of the measured torque.

The torque sensor of the flowmeter is a dual-range sensor; when the torque value decreases along the 1st range, from 0,5 Nm to 0,05 Nm, the measurement error progressively increases from 0.1% o.r. to 2% o.r.

The results show that the whole used range of the torque sensor is characterized by the ratio τ_{max}/τ_{min} = 0,5 Nm/0,0025 Nm = 200.

The accuracy of the used torque sensor determines the accuracy of the reaction flowmeter. In consequence, the measured errors of the reaction flowmeter repeat the variation of the dual-torque sensor, from 0,1% o.r. for $Q_{m_{max}}$ to 2% o.r. for $Q_{m_{min}}$, being considered $E_{F,\ max}$ = 2% o.r.

This variation of the measurement errors corresponds to a turndown of the reaction flowmeters of $Q_{m_{max}}/Q_{m_{min}} = (\tau_{max}/\tau_{min})^{\frac{1}{2}} = (200)^{\frac{1}{2}} = 14,14$, according to their functional equation (5.11).

The pressure drop Δp, depending on the measured mass flow rate $Q_{m_{max}}$, according to Table 5.2, is presented in Figure 5.5.

5.3.2.2.2 Main Resulting Features

The previous analysis demonstrates the following features of the respective reaction flowmeters:

TABLE 5.2

Comparison Between Experimental and Theoretical Calibration with Air

Torque τ	Inlet Parameters of Air				Pressure Drop Δp	Average Measured Flow Rate $Q_{m_{meas}}$	Calculated Flow Rate $Q_{m_{calc}}$	Matching $\frac{Q_{m_{calc}} - Q_{m_{meas}}}{Q_{m_{meas}}} \times 10^2$
	Pressure P	Temperature t	Humidity rH	Density ρ				
Nm	bara	°C	%	kg/m^3	bara	kg/h	kg/h	%
0,0025	0,9421	24,00	42,5	1,1030	0,0011	5,1120	5,1042	−0,15
0,0050	0,9432	23,67	42,7	1,1065	0,0020	7,2320	7,2298	−0,03
0,0100	0,9455	23,68	43,1	1,1110	0,0045	10,3740	10,2453	−1,24
0,0250	0,9520	24,12	42,6	1,1150	0,0110	16,5270	16,2284	−1,81
0,0500	0,9638	24,78	40,5	1,1270	0,0228	23,5340	23,0735	−1,96
0,1000	0,9855	26,51	35,5	1,1460	0,0445	33,5450	32,9048	−1,91
0,1500	1,0059	29,20	28,8	1,1580	0,0649	41,2950	40,5104	−1,90
0,2000	1,0286	32,81	24,4	1,1720	0,0876	47,9550	47,0593	−1,87
0,2500	1,0510	44,98	12,1	1,1500	0,1070	53,0620	52,1178	−1,78
0,3000	1,0690	44,99	12,2	1,1680	0,1280	58,4160	57,5372	−1,50
0,3500	1,0912	43,20	14,8	1,2010	0,1460	63,4000	63,0192	−0,60
0,3700	1,1008	34,10	20,7	1,2380	0,1575	66,0020	65,7852	−0,33
0,4200	1,1210	38,90	19,2	1,2500	0,1750	71,0020	70,4282	−0,81
0,4500	1,1277	40,02	18,5	1,3070	0,1910	74,7310	74,5437	−0,25
0,5000	1,1515	40,60	17,5	1,3320	0,2105	79,4190	79,3239	−0,12

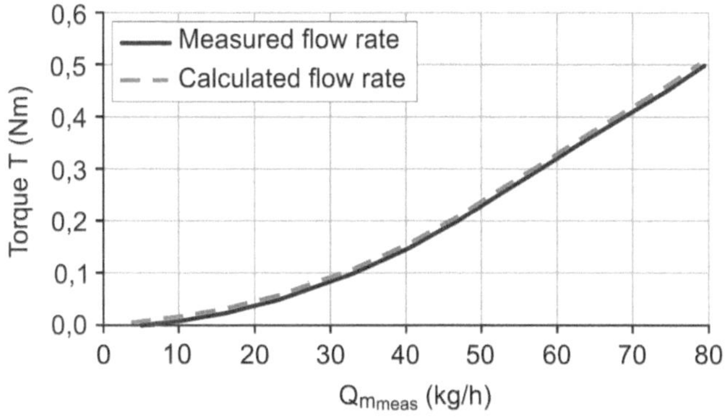

FIGURE 5.4
Mass flow rate of air depending on the measured torque (the reaction moment).

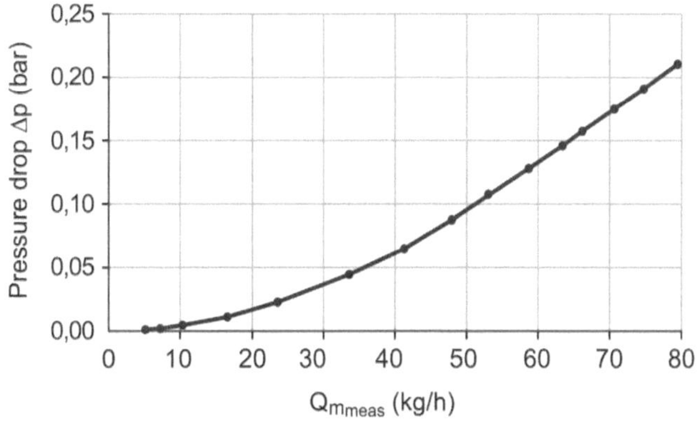

FIGURE 5.5
Pressure drop depending on the measured flow rate of air.

- This is a good matching of the experimental measured $Q_{m_{meas}}$ and theoretically calculated $Q_{m_{calc}}$. Their percentage differences are placed in a narrow band of the values – (0,03 … 1,96)%.
- The measured errors of the reaction flowmeter are placed in a range from 0,1% o.r., for $Q_{m_{max}}$, to 2% o.r. for $Q_{m_{min}}$.
- The turndown of the reaction flowmeters of $Q_{m_{max}}/Q_{m_{min}} = (\tau_{max}/\tau_{min})^{1/2} = (200)^{½} = 14,14$.
- The results show a moderate value of the pressure drop.

5.3.2.3 Preliminary Considerations on Potential Analytical Conversion of Flow Scales of Reaction Flowmeters from One Fluid to Another

The applying of the relation (5.3) for the measurement, by the same reaction flowmeter, of two different fluids with densities ρ_1 and ρ_2, ensures the theoretical possibility to establish an analytical correlation between the corresponding values of the mass flow rates Q_{m_1} and Q_{m_2}, when the related generated reaction forces have the same value, F_R.

In this respect, by dividing the relationships of Q_{m_1} and Q_{m_2}, the following equation is obtained:

$$Q_{m_1}/Q_{m_2} = (\rho_1/\rho_2)^{1/2} \tag{5.13}$$

Consequently, it is interesting to remark that is possible to determine the flow rate scale (corresponding to one fluid), by the analytical conversion of the flow rate scale of another fluid, which was experimentally established.

Thus, referring to the above presented results of calibration made by means of both water and air of the same type of reaction flowmeter, results show the following analytical dependence between the air mass flow rate $Q_{m_{air}}$ and the water mass flow rate $Q_{m_{water}}$.

Thus, comparative analysis of Tables 5.1 and 5.2, and using equation (5.11), results in the following relationship:

$$Q_{m_{air,conv}} = Q_{m_{water,meas}} \times \sqrt{\frac{\rho_{air}}{\rho_{water}}} \tag{5.14}$$

TABLE 5.3

Air Mass Flow Rate Computed by the Analytical Conversion of the Corresponding Experimental Measurement of the Water Mass Flow Rate

Moment of the Reaction Force	**Average Experimental Measured Flow Rate**		**Density**		$\sqrt{\frac{\rho_{air}}{\rho_{water}}}$	**Air Flow Rate Computed by Conversion**	**Matching Between Experimental and Conversion Calibrations**
	Water	**Air**	**Water**	**Air**			
τ		$Q_{m_{meas}}$	ρ_{water}	ρ_{air}		$Q_{m_{air,conv}}$	$\frac{Q_{m_{air,conv}} - Q_{m,meas}}{Q_{m,meas}} \times 10^2$
Nm	**Kg/h**		**Kg/m³**		**–**	**Kg/h**	**%**
0.010	308,869	10,374	998,20	1,111	0,033362	10,304488	−0,6701
0,025	486,290	16,527		1,115	0,033422	16,252784	−1,6592
0,150	1.199,002	41,295		1,158	0,034059	40,836809	−1,1096
0,250	1.549,122	53,062		1,150	0,033942	52,580299	−09078
0,350	1.830,906	63,400		1,201	0,034687	63,508636	+0,1714
0,450	2.076,891	74,731		1,307	0,036185	75,152301	+0,5638
0,500	2.189,455	79,419		1,332	0,036529	79,978600	+0,7046

Based on these results, it was elaborated the Table 5.3.

The variation range of the reaction moment τ, indicated in Table 5.3, renders for the same type of reaction flowmeter, the common zone of the variation ranges of the reaction moments generated at the measurement of water flow rate, respectively air flow rate.

Within this variation range of the reaction moment τ, are taken from Tables 5.1 and 5.2, for the same intermediate values of the moment τ, the corresponding values of the measured flows for water, respectively for air, as well as the afferent values of the measuring fluid density.

Based on the values of these parameters, and by using of the above indicated relationship, have been calculated the values of air flow rate $Q_{m_{air,conv}}$, by the analytical conversion of the values of water flow rates experimentally determined, corresponding to the same values of the reaction moment τ.

The last column of the Table 5.3 indicates a good matching between the values of air flow rate experimentally determined and those computed by the analytical conversion, inside of a narrow channel of (−1,6592 ... +0,7046)%.

These initial results provide a good basis for extending the study on the analytical conversion of flow scales to other fluids.

The range of Re numbers for the application of analytical flow conversion, corresponding to different fluid flow processes, is currently being studied.

5.4 Reaction Flowmeters with the Differential Measurement of the Pushing Force

5.4.1 Configuration and Operation

The second basic type of reaction flowmeters without moving parts is presented in Figure 5.6. It ensures the measuring of the reaction force by differential measurement of the output pushing force, (respectively pressure) produced by it, which is proportional to it. Figure 5.6 presents the longitudinal section of this type of flowmeter. It is no longer necessary to detail its cross section, because its casing is similar to that of the reaction flowmeter type, previously shown.

The measuring fluid enters the flowmeter through inlet connexion 1 ending with nozzle 2, which enters in nozzle 3 of the reaction tube 4.

The fluid flows through the reaction tube, and it is taken up by the convergent nozzle of the outlet connection 5, through which it is discharged from the flowmeter, without disturbance. The horizontal collinear connections 1 and 5 are rigid and sealed to flowmeter housing 6.

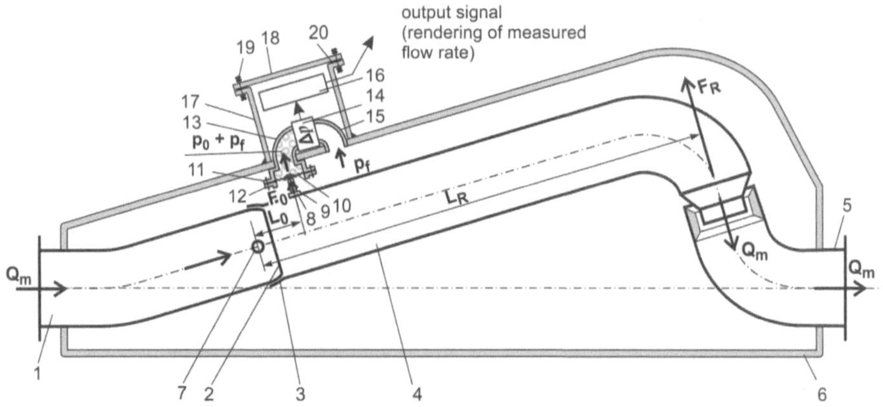

FIGURE 5.6
Reaction flowmeter with the differential measurement of output pushing force
Legend: 1 – immobile inlet connection, 2 – ending nozzle of inlet connection, 3 – inlet nozzle of reaction tube, 4 – reaction tube, 5 – outlet connection, 6 – housing of flowmeter, 7 – shaft, 8 – boss, 9 – pin, 10 – separation membrane, 11,12 – membrane tightening flanges, 13 – high pressure connection of Δp sensor, 14 – Δp sensor, 15 – low pressure connection of Δp sensor, 16 – electronic block, 17 – outer box housing, 18 – outer box cover, 19 – nut screw/system, 20 – gasket.

The radial equidistance and minimum friction between the nozzles 2 and 3 of the inlet connection 1 and the reaction tube 4 is made up by two very small bosses, placed vertically, up and down, around the vertical shaft 7, similarly to the solution presented in Figure 5.1b.

The rigorous positioning between the reaction tube and the inlet connexion is ensured by the rigorous concentric positioning of the nozzles 2 and 3, achieved by the vertical shaft 7.

The rigorous (x, y) positioning of the shaft 7 is accomplished by its passing through two holes placed on the upper and lower walls of the reaction tube support, which are rigorously located on the same vertical axis. The reaction tube 4 is stiffened with the shaft 7 that rests on its support, stiffened related to housing 6.

In the boss 8, at a distance L_0 from the shaft centre, the pin 9 is rigidly embedded, being in contact with the separation membrane 10 to which it permanently transmits the pushing force F_0 of the reaction tube 4, the force that is generated by the reaction force F_R.

Since then $L_R/L_0 > 1$, the F_0 force is amplified relative to the F_R reaction force.

The membrane 10 is tightly sealed between flanges 11 and 12 , and takes over the F_0 pushing force from pin 9. Consequently, on the side towards the fluid of the separation membrane 10 act, both the static pressure p_f of the fluid, and the pushing pressure $p_0 = F_0/A_m$, corresponding to F_0 pushing force (A_m being the active area of membrane).

The connection 13 (welded to flange 11 and stiffened against it by a nut, respectively rigidly and tightly mounted to the housing 6) is connected to the high pressure inlet (+) of the differential pressure sensor 14, and transmits to it the sum ($p_f + p_0$) of these pressures, provided by the transmission liquid, with which is previously fully filled the volume between the separation membrane and the sensing element of the sensor 14.

At the low pressure inlet connection of (–) of the sensor 14 is coupled the static pressure p_f of fluid, taken by the connection 15, which is rigid and seal-mounted on the housing.

Sensor 14 ensures a fully rigorous measurement of the differential pressure $\Delta p = (p_f + p_0) - p_f = p_0$, respectively of the reaction pressure, being structurally provided with the compensation function with the temperature and the pressure of the measuring fluid.

Thus, the sensor 14 indirectly measures the pushing force F_0, implicitly measures the reaction force F_R, and consequently measures the mass flow rate Q_m. The output signal of the sensor 14 is applied to the electronic block 16, where are previously stored both the calibration curve Q_m for the normal operating parameters of the measured fluid, and the facility of the flow rate compensation with pressure and temperature of fluid. Thus, the electronic block 16 renders the compensated value of the measured mass flow Q_m with pressure P and temperature T.

The differential pressure sensor 14, together with its connections 13 and 15, respectively electronic block 16, are placed inside an outer box that has its own housing 17, which is closed by the cover 18 with nut/ screw system 19, which tightens the gasket 20.

Since the pressure transmitting liquid between the separation membrane 10 and the differential pressure sensor 14 is practically incompressible, implicitly the displacement of membrane, and of the pin 9, is therefore almost null, to measure the entire range $Q_{m_{min}} \dots Q_{m_{max}}$. Correspondingly, the reaction tube displacement is extremely small, and *this type of flowmeter is practically a flowmeter without moving parts.*

5.4.2 Main Features

This type of the reaction flowmeters, using an industrial differential pressure sensor with a high accuracy of 0,02% o.f.s., can ensure the measuring of mass flow rate Q_m with the accuracy of (0,02...2) % o.r., for a turndown of $Q_{m_{max}}/Q_{m_{min}} = (\Delta p_{max}/\Delta p_{min})^{1/2} = 100^{1/2} = 10$, being considered $E_{F,\ max} = 2\%$ o.r.

It was obtained, for the laboratory conditions, an accuracy improvement of this sensor at 0,005%,o.f.s. that determined, for these conditions a turndown of flow measurement of $Q_{m_{max}}/Q_{m_{min}} = 400^{1/2} = 20$, for a corresponding accuracy range of (0.005 ... 2)% o.r.

The own microprocessor of the difference pressure sensor together the electronic block of the whole reaction flowmeter ensure the complete compensation of measured Q_m with temperature T and pressure P of the measured fluid.

These reaction flowmeters have the advantage that can be used for a wide range of fluids (liquids and gases), including the aggressive fluids, because the material of the differential pressure sensor, in contact with media, is stainless steel or viton.

5.5 Reaction Flowmeters with Direct Measurement of the Pushing Force

The third basic type of reaction flowmeters without moving parts ensures the measurement of the reaction force F_R, by the direct measurement of their output pushing force, which is proportional to F_R.

Further, for these basic type reaction flowmeters, there are successively presented two measurement variants of the output pushing force:

- by a load cell with strain gauge
- by its electromagnetic balancing

5.5.1 Reaction Flowmeters with Direct Measurement of Pushing Force by a Load Cell with Strain Gauge

5.5.1.1 Configuration and Operation

The configuration of this type of reaction flowmeters without moving parts is presented in Figure 5.7.

This basic type of reaction flowmeters is configured to ensure the direct measurement of its output pushing force F_0, which is proportional to the reaction force F_R.

The output pushing force F_0 is measured by a load cell with strain gauge.

Referring to the operation of this reaction flowmeter, the measuring fluid enters the flowmeter through the inlet connection 1, ending with nozzle 2, which enters in nozzle 3 of the reaction tube 4.

The fluid flows through the reaction tube 4 and respectively its convergent outlet nozzle 5, then it is taken up by the nozzle 6 of the outlet connection 7, being discharged from the flowmeter without disturbances.

The inlet 1 and outlet 7 are horizontal collinear connections and are rigidly and tightly fixed to the flowmeter housing 8.

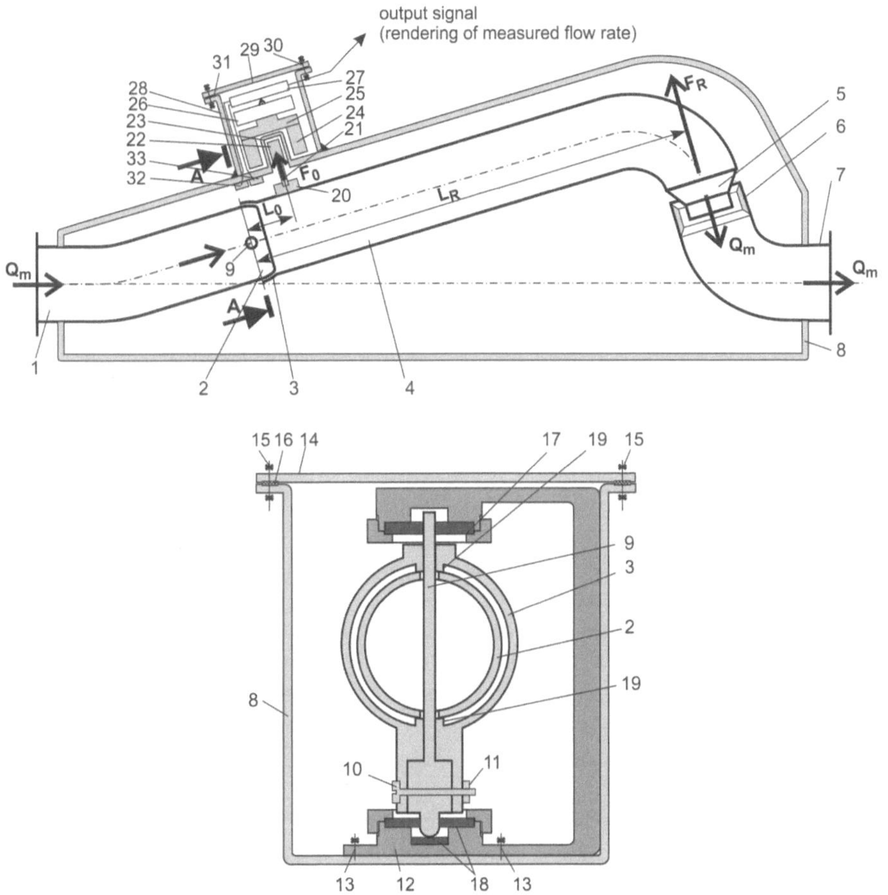

FIGURE 5.7
Reaction flowmeter with the output pushing force measurement by a load cell with strain gauge. a – longitudinal horizontal section, b – cross section with a plane A-A
Legend: 1 – inlet connection, 2 – ending nozzle of inlet connection, 3 – inlet nozzle of reaction tube, 4 – reaction tube, 5 – outlet nozzle of reaction tube, 6 – nozzle of outlet connection, 7 – outlet connection, 8 – flowmeter housing, 9 – shaft, 10 – screw, 11 – nut, 12 – support of flowmeter, 13 – screw, 14 – flowmeter cover, 15 – nut /screw system, 16 – gasket, 17 – upper bearing, 18 – bottom bearing, 19 – inner boss, 20 – outer boss, 21 – pin, 22 – inner magnet, 23 – closed tube, 24 – outer ring magnet, 25 – magnet support, 26 – force sensor, 27 – electronic block, 28 – housing of outer box, 29 – cover of outer box, 30 – nut /screw system, 31 – gasket, 32 and 33 – sensors of temperature and pressure.

The reaction tube is provided with the shaft 9 perpendicular to it and rigid with it, by fixing with the screw 10 and the nut 11. The shaft 9 rests on its support 12, which is stiffened related to housing 8 by screws 13. The housing 8 is closed by flowmeter cover 14, with nut- screw system 15, which tightens the gasket 16.

The rigorous positioning of the shaft 9 ensures on the one hand, the rigorous radial concentric equidistance between the nozzles 2 and 3, and on the other hand the insignificance of its friction with upper bearing 17 and the bottom bearings 18.

The equidistance between the nozzles 2 and 3 is constructively achieved by two small inner bosses 19 of the nozzle 3 of the reaction tube, placed up and down around the vertical shaft 9. Thus, it is ensured, on the one hand, an insignificant friction between nozzles 2 and 3, that have a rigorous concentric positioning, and on the other hand, the potential rotating mobility of the reaction tube 4, around the vertical axis of the shaft 9 without any inconvenience.

The reaction tube 4 is provided with the outer boss 20. In this boss, at a distance L_0 from the shaft axis, the pin 21 is rigidly embedded.

On the end of pin 21 is rigidly fixed the magnet 22, which is placed inside a small closed tube 23.

The ensemble (pin 21 and magnet 22) can move freely in relation to the closed tube 23.

The tube 23 is fixed sealed by welding to housing 8.

Outside the tube 23, around it, is placed a ring magnet 24 with a free displacement in relation to the tube 23, having the inner hole with a suitable greater diameter to the outer diameter of it.

Due to the magnetic coupling between the magnets 23 and 24, any value variation of the force F_0, is followed by the tendency of the pin 21 movement, (respectively of inner magnet 22), that determines the tendency of movement of the outer magnet 24.

The magnet 24 is rigidly fixed to its support 25. The support is constructively fixed on the mobile part of the force sensor 26, and permanently transmits to it the pushing force, received by the magnet 24 (respectively by the inner magnet 22), with the instantaneous value of F_0.

According to its principle of operation, the force sensor 26 (a load cell with strain gauge), receiving the pushing force F_0, to any very small tendency of displacement of its mobile part, (corresponding to any small variation of the force F_0), ensures its measurement.

The electrical output signal of sensor 26, being proportional to the value of measured force F_0, is applied to the electronic computing block 27.

The outer ring magnet 24 (together with its support 25, the force sensor 26, and the flow rate computing block 27) are properly positioned inside a box, having its own housing 28, which is closed by the cover 29 with nut-screw system 30, that tightens the gasket 31.

The housing 28, evidently being not in contact with the measured fluid, is rigidly fixed (e.g., by welding) to the housing 8, which contrary, is permanently in contact with the fluid.

The force sensor 26 is rigidly fixed to the housing 28, in a rigorous position, to be ensured its properly functional positioning in relation to the pushing force pick-up system (constituted by the couple of the two magnets 22 and 24).

The block 27, according to the previously known analytical relationship between forces F_R and F_0, computes the corresponding value of the reaction force F_R.

On this base, the block 27, where was previously stored the calibration curve $Q_m = Q_m(F_R)$, ensures, according to the measured value of F_R, the calculation of the instantaneous value of the compensated mass flow rate Q_m with T and P, which are measured by the sensors 32 and 33.

Thus, the electrical output signal of the electronic block 27 is proportional to the instantaneous value of the respective measured mass flow rate Q_m.

The necessary displacement of the mobile part of sensor 26, and implicitly of the pin 21, respectively of the reaction tube 4, is very small, for the measurement of entire range $Q_{m_{min}} \ldots Q_{m_{max}}$ of the reaction flowmeter.

Consequently *this type of reaction flowmeter is practically a flowmeter without moving parts.*

5.5.1.2 Main Features

Using a force sensor consisting of a load cell with strain gauge, realised with a high accuracy of 0,014% o.f.s. compensated with the temperature, it was achieved a type of reaction flowmeter that ensures the measurement of the mass flow rate Q_m with a turndown of $Q_{m_{max}}/Q_{m_{min}} = (2/0{,}014)^{1/2} = 11{,}95$ for a corresponding accuracy range of (0,014 ... 2)% o.r and $E_{F,\ max} = 2\%$ o.r.

5.5.2 Reaction Flowmeters with Direct Measurement of Pushing Force by its Electromagnetic Balancing

5.5.2.1 Configuration and Operation

These reaction flowmeters have the same configuration as shown in Figure 5.7.

The only difference is that, for the measurement of the output pushing force F_0 is used, instead of a force sensor achieved by a load cell with strain gauge (according to above variant), one with the electromagnetic balancing of this force.

Consequently, for this variant, item 26 from Figure 5.7 is ensured by a load cell with the measurement of the output pushing force F_0 by its electromagnetic balancing.

For uses at low ambient temperatures it is necessary to be included in this configuration, supplementary, an ambient temperature sensor inside the box delimited by the housing 28, to be corrected, with temperature, the information given by the force sensor.

5.5.2.2 Main Features

This second variant ensures a very important improvement of the performance of "reaction flowmeters with direct measurement of output pushing force".

In this respect, by using of a force sensor with the force measurement by its electromagnetic balancing, having a high accuracy of 0,0003% o.f.s., is achieved the mass flow rate Q_m measurement, with a potential turndown of $Q_{m_{max}}/Q_{m_{min}} = (2/0,0003)^{1/2} = 81,65$, for a corresponding accuracy range of (0,0003 … 2) % o.r, being considered $E_{F,\ max} = 2\%$ o.r.

6

Reaction Flowmeters Without Moving Parts and Linear Dependence "Flow Rate – Measured Characteristic Parameter"

6.1 Configuration and Operation

Basic configuration of this type of reaction flowmeter without moving parts is presented in Figure 6.1.

Fluid having the mass flow rate Q_m, enters the flowmeter through inlet connection 1, passes through reaction tube 2 (articulated with inlet connection 1 by vertical inner shaft 3), and then is discharged from the flowmeter through outlet connection 4.

Connections 1 and 4 are fixed rigidly and tightly to housing 5.

Corresponding to the value of the mass flow rate Q_m, is generated and acts, along the axis of outlet elbow of reaction tube 2, at a distance L_R from the axis of inner shaft 3, the reaction force F_R, which determines the reaction moment M_R, related to the symmetry axis of inner shaft 3.

Corresponding to the reaction force F_R, the reaction moment M_R tends to rotate reaction tube 2 around the axis of inner shaft 3, with an angle α, proportional to reaction force F_R.

Inner shaft 3, being rigid with reaction tube 2, tends to rotate with the same angle α, together with inner magnet 6, which, in its turn, is stiffened against inner shaft 3 by sleeve 7.

In its turn, magnet 6 inside housing 5 of the flowmeter determines, by magnetic coupling, rotation with same angle α of outer magnet 8, which is placed outside of housing 5, and rigidly fixed, by strong gluing, to the profiled sleeve of outer shaft 10, that is rigidly fixed to the inner ring of bearing 9.

On the other hand, the outer ring of bearing 9 is rigidly fixed on its support 11, which in turn is rigidly fixed on housing 5.

Also, outer shaft 10 is fixed rigidly to slider rod 12, which thus rotates with the same angle α as outer shaft 10 and as outer magnet 8.

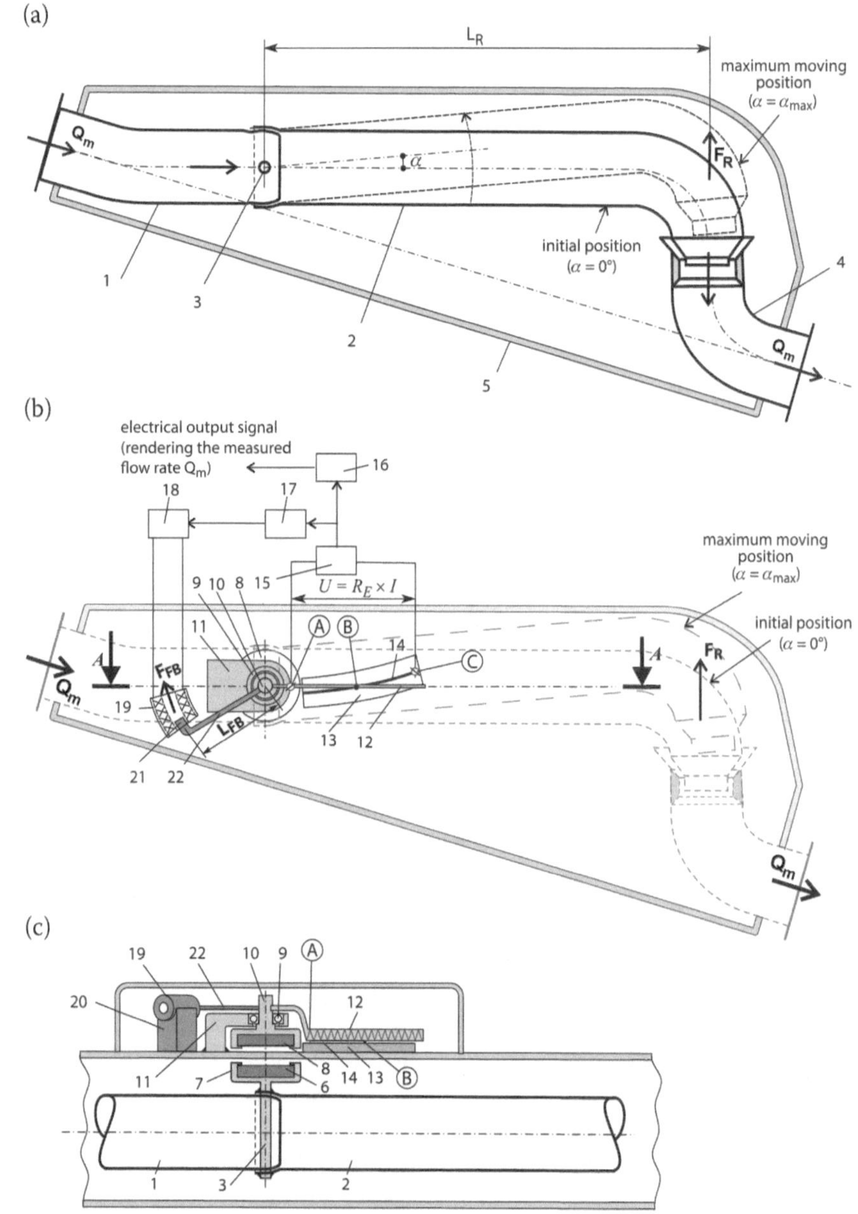

FIGURE 6.1
Configuration of the reaction flowmeters without moving parts and linear dependence "flow rate-measured characteristic parameter". a – horizontal longitudinal section, b – top view, c – cross section with plane A-A
Legend: 1 – inlet connection, 2 – reaction tube, 3 – inner shaft, 4 – outlet connection, 5 – housing, 6 – inner magnet, 7 – sleeve, 8 – outer magnet, 9 – bearing, 10 – outer shaft, 11 – bearing support, 12 – slider rod, 13 – nonlinear profiled cam, 14 – electrical contact conductor, 15 – voltage transducer, 16 – electronic output block, 17 – calculation block of U^2, 18 – feedback block, 19 – feedback coil, 20 –coil support, 21 – ferromagnetic part, 22 – lever.

The core of slider rod 12 is made of an electrical insulating material on which an electrical resistance R_E is fixed along the entire length of its horizontal profile.

This electrical resistance R_E is achieved by a continuous uniform spirally winding of an wire, with a good mechanical resilience and with a high electrical resistivity that varies very little with ambient temperature.

Under the horizontal plane of the angular rotation of slider rod 12 is positioned cam 13 in a horizontal plane, which is rigidly fixed on the outer side of housing 5.

The contour of cam 13 is profiled thus as the variation ΔR_C of length of its radii to respect the relationship $\Delta R_C = c \times \alpha^{1/2}$, where c is a constant and α is the value of the rotation angle of slider rod 12.

On the profiled contour of cam 13 is rigidly fixed conductor 14 with good conductance.

Cam 13 is positioned so that the origin of its radii is placed on the rotation vertical axis of outer shaft 10.

Through the electrical resistance, which is uniform wound along its length of its horizontal profile, slider rod 12 is in permanent contact, noted B, with electrical conductor 14, which is rigidly fixed on the contour of cam 13.

Position of contact B is variable, corresponding to the angular position α of slider rod 12.

Thus, by its rotation with angle α, slider rod 12 has the functional role to ensure a variable electrical resistance R_E, depending on the profile of cam 13, because its functional length varies similarly with the length R_i of the cam radius, corresponding to position of contact point B.

Its functional length corresponds to the distance between its first end, noted A (which permanently is placed at a constant distance to the rotation axis of slider rod), and its second end, noted B, positioned at variable distance to the its rotation axis (depending on the slider rod rotation angle α), the end B being rendered by its electrical contact point with electrical conductor 14, which is fixed on the contour of the cam.

The end B of the electrical resistance R_E is a mobile point, ensured by the electrical contact point between mobile slider rod 12 and electrical conductor 14.

It results that the value of the electrical resistance R_E permanently renders the length variation of the cam radius, respectively its dependence on square root of the rotation angle α, thus R_E being proportional to the square root of force F_R.

Being supplied with a constant electric current of a very small intensity, the variable electrical resistance R_E, determines a variable voltage difference U between its ends A and B (respectively C), which is measured by transducer 15.

Voltage U represents the characteristic parameter of this basic type of reaction flowmeter, and consequently its measurement determines the value of the output signal of the reaction flowmeter, which renders the measured flow rate Q_m.

Voltage U, being proportional to the electrical resistance R_E, is thus proportional to the square root of the reaction force F_R, according to the relationship $U = c_0 \times F_R^{1/2}$.

Consequently, due to, on the one hand, this nonlinear dependence between characteristic parameter U and F_R, and, on the other hand, the nonlinear dependence between F_R and Q_m according to equation (3.3), "the linear dependence between the measured mass flow rate Q_m and the voltage U, the characteristic parameter of the reaction flowmeter" is obtained.

Accordingly, the output signal of voltage transducer 15 is received by the output electronic block 16, that by using the relationship between U and Q_m, previously stored, computes the value of the measured mass flow rate Q_m, and renders it by its output electrical signal (that constitutes the parameter Q_R of rendering the measured flow rate Q_m).

In addition, the output signal of voltage transducer 15 is also received by electronic calculation block 17 of *the* U^2 value, that gives at its output a signal proportional to the value of reaction force F_R, the signal which is applied to the feedback block 18.

In its turn, block 18 generates an electrical output signal, with the value of the current intensity I proportional to its input signal, which is applied to feedback coil 19. Coil 19 is rigidly fixed to housing 5 by support 20.

Thus, coil 19 is permanently supplied by an electrical current with the intensity I (proportional to the value of the reaction force F_R) and generates an electromotive force (named "feedback force") F_{FB}.

This feedback force F_B, being proportional to the current intensity I (respectively to the reaction force F_R), acts on ferromagnetic part 21, which is fixed rigidly perpendicular to lever 22.

The attraction effect of the force F_{FB}, applied on part 21, determines the feedback moment $M_{FB} = F_{FB} \times L_{FB}$, which acts on outer shaft 19.

Thus, the moment M_{FB} is proportional to reaction force F_R and tends to rotate outer shaft 10 in an opposite sense to the rotation action determined by the moment M_R of the reaction force F_R, and permanently tends to balance it.

Consequently, due to this permanent feedback moment (which is a variable resistance moment developed by coil 19, opposite to reaction moment M_R), the rotation of outer shaft 10 (which repeats the rotation of reaction tube 2) stops at the equilibrium value of the rotation angle α, when the feedback moment M_{FB} equals the reaction moment M_R.

Remark: Because the maximum functional rotation angle α of the reaction tube has a small value up to 8° .... 9°, *practically these flowmeters are considered reaction flowmeters without moving parts.*

In conclusion, for an instantaneous value of the measured mass flow rate Q_m, a reaction force F_R is generated, which, by balancing its moment

M_R with the feedback moment M_{FB}, determines a proportional equilibrium value of rotation angle α, of reaction tube 2, implicitly of outer shaft 10 (respectively of slider rod 12), which is nonlinearly converted to an electrical voltage U having the expression ($U = c \times \alpha^{1/2} = c_0 \times F_R^{1/2} = C \times Q_m$), that is measured, and obtained by electronic block 16 the output signal of reaction flowmeter.

The voltage U expresses the characteristic parameter of this basic type of reaction flowmeter.

Thus, F_R having a nonlinear dependence to mass flow rate Q_m, by replacing its analytical equation in the above expression of U, results in linear dependence between the measured mass flow rate Q_m (input parameter of reaction flowmeter) and the voltage U (the characteristic parameter of the reaction flowmeter).

6.2 Functional Equation

According to the configuration of the reaction flowmeter shown above, on the one hand, the outer shaft 10 permanently renders the rotation angle α of the reaction tube 2. On the other hand, outer shaft 10 is actuated by two mentioned opposite moments of forces, respectively:

- on the one hand, by the reaction moment M_R of the reaction force F_R, expressed by the analytical relationship:

$$M_R = F_R \times L_R \tag{6.1}$$

- on the other, by the feedback moment M_{FB} expressed by the analytical relationship:

$$M_{FB} = F_{FB} \times L_{FB} \tag{6.2}$$

As a result, the angle of rotation α of outer shaft 10 permanently renders the angular position (respectively of slider rod 12), which ensures the equilibrium of these two opposite moments, respectively the feedback moment M_{FB}, which equals the reaction moment M_R.

The following are presented successively, from the analytical point of view:

- the effect of the reaction moment M_R
- the feedback ensured by the configuration of the flowmeter to the effect of M_R, by opposing the moment M_{FB}

The reaction moment M_R determines the rotation of slider rod 12 with an angle α, that repeats the rotation of outer shaft 10.

Slider rod 12 slides on cam 13, which has its contour profiled according to the requirement that the change ΔR_C of the length of its radii, on which the resistive rod 12 slides, depends on the rotation angle α of outer shaft 10 (and of slider rod 12) according to relationship:

$$\Delta R_c = c_1 \times \alpha^{1/2} \tag{6.3}$$

where:
c_1 – constant

In this respect, the length of cam radius, R_C, has the expression:

$$R_C = R_{C_0} + \Delta R_C \tag{6.4}$$

where:
R_{C_0} – the rod length (radius), for α = 0, from which the effect of the electrical resistance begins.

This angle is converted into an electrical resistance R_E (respectively an electrical voltage U, measured by the transducer 15 between the ends A and B of the electrical resistance R_E, which is supplied with a very small electrical current), corresponding to length ΔR_C, having the following analytical expression:

$$U = I_0 \times R_E = I_0 \times c_2 \times \Delta R_C = c_3 \times \alpha^{1/2} \tag{6.5}$$

where:
I_0 – very small electrical current intensity – constant
$c_3 = I_0 \times c_2 \times c_1$ – constant

The output information of transducer 15 is then processed by block 17 that calculates the value of U^2, and applied to block 18. In turn, block 18 generates an electrical signal with current intensity $I = c_4 \times U^2$, that supplies coil 19. Coil 19 accordingly develops, proportional to it, an electromotive force F_{FB}, which has the analytical expression:

$$F_{FB} = c_5 \times I = c_5 \times c_4 \times U^2 = c_5 \times c_4 \times c_3^2 \times \alpha = c_6 \times \alpha \tag{6.6}$$

where:
$c_5 \times c_4 \times c_3^2 = c_6$ – constant

The feedback force F_{FB} determines the feedback moment M_{FB}, that actuates ferromagnetic part 21, respectivey outer shaft 10 by lever 22, in the opposite sense to the reaction moment M_R, and has the following analytical expression:

$$M_{FB} = F_{FB} \times L_{FB} = L_{FB} \times c_6 \times \alpha = c_7 \times \alpha \qquad (6.7)$$

where:
$L_{FB} \times c_6 \times c_5 = c_7$ – constant

According to the above explanations, the angle of rotation α of outer shaft 10 permanently renders the angular position (and of slider rod 12) which ensures the equilibrium of the two opposite moments (when the feedback moment M_{FB} equals the reaction moment M_R), expressed by the equation:

$$M_{FB} = M_R \qquad (6.8)$$

By replacing of M_{FB} and M_R and by their processing, it results in the following:

$$\alpha = L_R \times F_R \times c_7^{-1} \qquad (6.9)$$

According to Chapter 6.1, the characteristic parameter of this basic type of reaction flowmeter is the electrical voltage U, which is measured by transducer 15 and is converted by electronic block 16 in a proportional electrical output signal.

By replacing the above expression of the angle α, in the expression (6.5) it results the following equation of the voltage U:

$$U = c_3 \times (L_R \times F_R \times c_7^{-1})^{1/2} = c_8 \times F_R^{1/2} \qquad (6.10)$$

where:
c_8 – constant

On the other hand, by replacing F_R, according to its dependence by the measured mass flow rate Q_m, indicated in equation (3.3), it results the linear dependence between the characteristic parameter U and the input parameter Q_m, for this basic type of reaction flowmeter.

Consequently, *this specific basic type of reaction flowmeter has the following linear functional equation:*

$$U = c_8 \times (k_1 \times \rho^{-1})^{1/2} \times Q_m = c \times \rho^{-1/2} \times Q_m \qquad (6.11)$$

where:
$c_8 \times k_1^{1/2}$ – constant

The above relationship renders "the linear dependence between the measured flow rate Q_m and the characteristic parameter (voltage U) of the reaction flowmeters".

6.3 Main Features

This basic type of reaction flowmeter, corresponding to its specific configuration, operation, and functional equation presented above, has the main feature to ensure the mass flow rate measurement with a significant increase of turndown value, correlated with an optimum variation range of accuracy (expressed as % o.r.).

The increase of the turndown value is determined by the linear dependence "flow rate – measured characteristic parameter (voltage U)". This linear dependence is obtained by the conversion of the reaction force F_R, without its measurement and processing, into the characteristic parameter (voltage U) of the reaction flowmeter, proportional to the square root of F_R. This conversion is only constructively achieved by the functioning of its specific configuration.

Further, we present the considerations that ensure the determination of the analytical expression of the turndown for this basic type of reaction flowmeter.

In this respect, initially it is necessary to remind the analytical expression (5.11) of the turndown, previously established for the reaction flowmeters without moving parts and nonlinear dependence "flow rate – measured characteristic parameter".

Consequently, for all these types of the reaction flowmeters, their nonlinear functional and analytical dependence between F_R and Q_m, expressed by the equation (3.3), determines the nonlinear dependence between the values of the two essential characteristics of them, respectively the turndown T_F and the maximum permissible measurement error of the reaction flowmeters, expressed as $E_{F_{max}}$,(% $o.r.$).

Starting from the above analysis results in obtaining a linear dependence between the measured flow rate Q_m and the effect of the reaction force F_R (the voltage U that becomes "the characteristic parameter" of the reaction flowmeter, being consequently measured instead of the reaction force).

In turn, the essential effect of this achievement is the obtaining of the linear dependence between the turndown T_L and $E_{F_{max}}$ (% $o.r.$), according to the relationship:

$$T_L = \frac{Q_{m_{max}}}{Q_{m_{min}}} = \frac{E_{F_{max}}(\%\ o.r)}{E_S(\%\ o.f.s)} \tag{6.12}$$

TABLE 6.1

Theoretical Turndown of the Reaction Flowmeters without Moving Parts and Linear Dependence "Flow Rate – Characteristic Parameter"

E_S	$E_{F_{max}}$	T_L
% o.f.s.	**% o.r.**	–
0,001	2	1.999,87
	1,5	1.500,00
	1	999,82
	0,5	499,96
	0,1	100,00

The comparison of this relationship with the previous one indicates a significant growth of the *T* turndown value with reference to the same $E_{F_{max}}$ (% *o.r*) value.

In this respect, Table 6.1 presents, as a case example, the values of turndown, theoretically determined, corresponding to the different values of the maximum permissible errors $E_{F_{max}}$, which can be ensured by the reaction flowmeters configured above, using a voltage transducer with the accuracy E_S = 0.001% o.f.s.

For each value of $E_{F_{max}}$ indicated in Table 6.1, the reaction flowmeter accuracy varies along the ranges limited by the value 0,001% o.r., corresponding to $Q_{m_{max}}$, and the value of $E_{F_{max}}$ (% o.r.), corresponding to $Q_{m_{min}}$, this limit being previously established for the measurement of the mass flow rate Q_m.

7

Reaction Flowmeters with Moving Parts

7.1 Basic Reaction Measurement System Adapted to Reaction Flowmeters with Moving Parts

Figure 7.1 presents the specific configuration of the reaction measurement system, of all reaction flowmeters with moving parts, obtained by adapting the basic system configuration presented in Figure 3.1.

The specific and common "reaction measurement system" of all reaction flowmeters with moving parts consists of the pair of immobile inlet tube 1, configured as an extension of the inlet connection of flowmeter and the reaction element 2, RE, provided with two fluid discharge nozzles, outlet 1 and outlet 2, placed diametrically opposite.

The reaction element 2 is provided also with central shaft 3, perpendicular to it and rigid with this, by fixing with screw 5 and nut 6. Thus, shaft 3 ensures, together with upper bearing 4 and bottom bearing 7, the potential rotation of RE about its vertical symmetry axis.

"The reaction element" (RE) is symmetrically configured, by connecting two identical reaction tubes, with the same configuration, previously presented in Figure 3.1 for "the reaction tube", the main part of the reaction measurement system, specific for the reaction flowmeters without moving parts. Specifically for the reaction flowmeters with moving parts is the placement of these two reaction tubes with their outlets in opposite directions.

Both mandatory requirements, previous mentioned in Chapter 3, are met by each component reaction tube of RE, individually and in consequence by "the reaction element" as a whole.

This specific configuration of the reaction element (RE) of the reaction flowmeters with moving parts ensures the possibility to detect the driving (reaction) torque τ_D, developed by the pair of the two opposite reaction forces (respectively F_{Ry}), that rotate the RE about its central vertical axis, together with shaft 3. Starting from these bases (configuration and operating

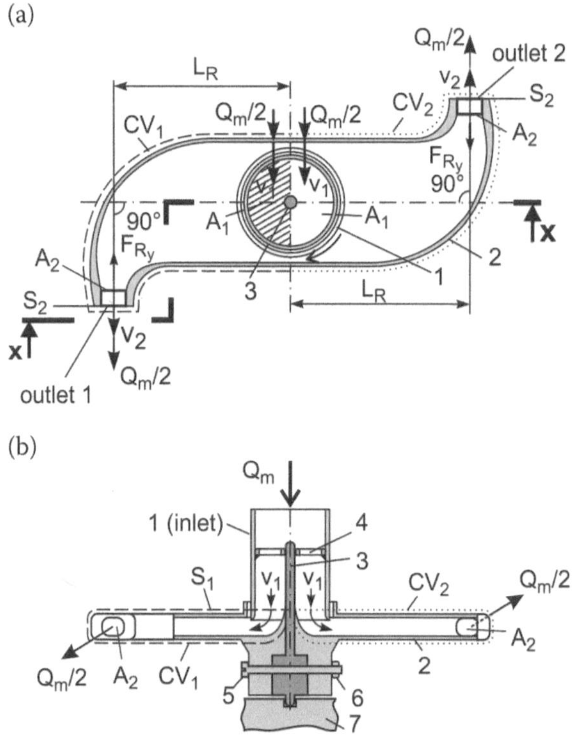

FIGURE 7.1
Configuration and operating principle of the basic reaction measurement system, adapted to reaction flowmeters with moving parts a. longitudinal section, b. cross section in steps (x-x). Legend: 1 – immobile inlet tube, 2 – reaction element, 3 – central shaft of rotation, 4 – upper bearing, 5 – screw, 6 – nut, 7 – bottom bearing.

principle) are determined the functional equations of the reaction flowmeters with moving parts.

7.2 Functional Equations

The analysis of the specific reaction measurement system of the reaction flowmeters with moving parts is ensured according to its configuration presented in Figure 7.1, related to two reference control volumes CV1 and CV2 of the two component reaction tubes of the RE of the reaction flowmeter.

The control volumes CV1 and CV2 are similarly configured with "the reference control volume" CV of the reaction tube presented in Figure 3.1, for the reaction flowmeters without moving parts.

In this respect each of these control volumes CV1 and CV2 has the faces at the inlet (in the plane of section S1, achieved by its corresponding half of inlet connection 1 area, shaded for CV1 and not shaded for CV2, with inner area A1), and at the outlet (in the plane of section S2), with inner area A2 of its outlet (outlet 1 for CV1 and outlet 2 for CV2), each encompassing the walls of its corresponding reaction tube. This "rotary reaction element", RE, being constituted by the connection of the two identical reaction tubes, with the configuration previously presented in Figure 3.1, the driving (reaction) force developed by each reaction tube, component of RE, will follow the similar expression as in equation 3.3, with the input flow rate $Q_m/2$ for each component reaction tube, instead of Q_m, as in Figure 3.1.

7.2.1 Flowmeter Factor Equation

For the characterization of the calibration curve of the reaction flowmeters with moving parts, K flowmeter factor is used, with the general significance:

$$K = \omega / Q_m \tag{7.1}$$

where:

ω – angular velocity of RE

To establish the functional equation of K factor, two significant cases of RE operation are analyzed successively: in ideal conditions and in real conditions.

A. Operation in ideal conditions

This operation case corresponds to hypothetical RE operation (where there are no forces acting to slow down the RE, respectively retarding torques due to fluid drag on the RE and drag in RE bearings).

For "ideal conditions" the following "ideal parameters" correspond and will be used:

$$K_{id} = \omega_{id} / Q_m \tag{7.2}$$

On the other hand, because $v_{output,\ ideal} = v_{2.id}$ is the RE tangential velocity (implicitly of the output fluid velocity) at radius L_R, it results in the following expression of the ideal angular velocity:

$$\omega_{id} = v_{2.id} / L_R \tag{7.3}$$

where:

L_R – distance between the symmetry axis of the outlet and the vertical symmetry axis of RE (its rotation axis).

By processing of the previous equations, and because 0,5 $Q_m = v_{2.id} \times A_2 \times \rho$, it results in the following:

$$K_{id} = \omega_{id} / Q_m = \frac{1}{2 \times A_2 \times L_R \times \rho} = K_0 \times \rho^{-1} \tag{7.4}$$

where:

A_2 – inner area of outlet 1 and of outlet 2
ρ – fluid density
K_0 – constructive reaction flowmeter factor

The constructive reaction flowmeter factor K_0 has the expression:

$$K_0 = (2 \times A_2 \times L_R)^{-1} \tag{7.5}$$

It is important to remark that the K_0 factor is the specific constructive factor of each type of the reaction flowmeter with moving parts, determined only from the geometry of the reaction measurement system.

We observed a complete similitude between RE "ideal operation" of the reaction flowmeters with moving parts, under the current analysis, and the "real operation" of the reaction tube of the reaction flowmeters without moving parts, presented in Chapter 5.2.1 because for both there are no forces acting to drag RE, respectively the reaction tube.

Consequently, because the ideal driving force $F_{d,id}$ (exerted on RE) is itself the reaction force $F_{R_{y,id}}$ (exerted by fluid on RE), it results that $F_{d,id} = F_{R_{y,id}}$, which has the same expression as in equation (3.3).

By replacing of the input flow Q_m with $Q_m/2$, and by an adequate processing of the equation (3.3), related to v_2, the following equation results:

$$F_{d,id} = F_{R_y,id} = 0{,}25 \times k_1 \times Q_m^2 \times \rho^{-1} \tag{7.6}$$

where:

$k_1 = k_1\ (A_1, A_2)$ – constructive constant

B. Operation in real conditions

In the situation of real operation, there are forces acting to slow down RE, that cause the retarding torques due to fluid drag on the RE and drag in RE bearings.

The change in momentum of the fluid crossing the RE inlet and outlet planes S_1 and S_2 is the driving force opposing the retarding toques, and the movement equation has the expression:

$$I \times \frac{d\omega}{dt} = \tau_d - \tau_r \tag{7.7}$$

where:
I – moment of inertia of RE
t – time

In the momentum approach to the real physical model of "the reaction flowmeter with mobile parts", the steady-state angular velocity of the reaction element, RE, satisfies the equilibrium $\tau_d = \tau_r$ of the driving torque τ_d and the retarding torque τ_r on the RE, because $\frac{d\omega}{dt} = 0$ in the movement equation of RE.

It is applied the conservation of momentum to the control volume containing the reaction element and neglecting pressure effects.

In this "real conditions", a decrease of both the output tangential velocity, from ideal value $v_{2,\mathrm{id}}$ to real value v_2 (respectively the corresponding angular velocity, from ω_{id} to ω), and the driving (the reaction forces) from $F_{d,id} = F_{R_{y,id}}$ to $F_d = F_{R_y}$, due to the total flow-induced drag force F_r exerted on RE, is expressed by the equation:

$$F_r = F_{d,id} - F_d = F_{R_{y,id}} - F_{R_y} = 0{,}5 \times Q_m \times L_R \times (\omega_{id} - \omega) \tag{7.8}$$

Consequently the retarding torque has the expression:

$$\tau_r = L_R^2 \times Q_m \times (\omega_{id} - \omega) \tag{7.9}$$

By dividing with Q_m and by processing, it results in the following:

$$\frac{\omega}{Q_m} = \frac{\omega_{id}}{Q_m} - \tau_r/(L_R^2 \times Q_m^2) \tag{7.10}$$

By replacing $\frac{\omega_{id}}{Q_m}$ from expression (7.4), the general expression of K flowmeter factor for "real conditions", is obtained:

$$K = \frac{\omega}{Q_m} = (2 \times A_2 \times L_R \times \rho)^{-1} - \tau_r/(L_R^2 \times Q_m^2) \tag{7.11}$$

7.2.2 Retarding Torque Equation

The total retarding torque τ_R is the sum of three specific types of retarding torques, as follows:

$$\tau_r = \tau_f + \tau_m + \tau_s \tag{7.12}$$

where:

τ_f– fluid friction torque (the sum of the fluid viscous drag torque exerted by the fluid on all moving surfaces of rotary RE and in its bearings, including tip clearance).

τ_m – mechanical friction drag torque of bearings (it is considered to be constant, independently of the rotation velocity ω).

τ_s – rotation sensor drag torque (e.g., the magnetic pickoff torque for the reaction flowmeters equipped with a magnetic pickoff).

It is continued with the synthetic successive presentation of each specific component of the total retarding torque τ_r.

A. *The friction (viscous) drag torque* τ_f.

This torque is exerted by the metered fluid and acts both on all moving RE surfaces including within its bearing, respectively in the space between the RE tip and the flowmeter housing, being the sum of the specific torques, presented below.

- Fluid drag torque on RE surface.

This drag torque has with the expression:

$$\tau_{RE_f} = 0,5 \times C_1 \times C_D(Re) \times A_{RE} \times \rho \times \omega^2 \tag{7.13}$$

where:

C_1 – RE constructive coefficient,

$C_D(Re)$ – fluid drag coefficient that is a function of Reynolds number

A_{RE} – RE surface area

Remark: Other fluid drag torque, of a small value as compared to τ_{RE_f} torque, is the tip clearance fluid drag of forces, caused by the flow around the tip of reaction element RE. It depends especially on Reynolds number, and respectively on the size of clearance, type of clearance (enclosed or not enclosed in the housing), the RE shape, and its dimensions.

This drag torque tends to influence the characteristic curve of reaction flowmeter for its lower zone, of the transition from laminar to turbulent flow.

- Fluid drag torque in the bearing.

This drag torque has the expression:

$$\tau_{Bf} = C_{B1} \times \rho \times v \times \omega \tag{7.14}$$

where:
C_{B1} – constant
v – cinematic viscosity of fluid

B. *Mechanical retarding torque.*

It is constituted by the sum of the following torques:

$$\tau_m = \tau_{Bm} + \tau_{Bat+di} \tag{7.15}$$

The components of the mechanical retarding torque τ_m have the following expressions:

- Vertical bearing friction retarding torque with the expression:

$$\tau_{Bm} = C_{B2} \times R_b \times F = C_{B2} \times R_b \times \left[G_{RE} + \frac{Q_m^2}{2 \times \rho \times A_{inlet}} \right] \tag{7.16}$$

where:
C_{B2} – constant
R_b – radius of bearing
F – vertical force exerted on the vertical shaft of the bearing
G_{RE} – reaction element RE weight

- Retarding torque due to axial thrust and dynamic imbalance, with the expression:

$$\tau_{Bat+di} = C_{B3} \times \omega^2 \tag{7.17}$$

where:
C_{B3} – constant

C. *Rotation sensor drag torque,* with the expression:

$$\tau_s = C_s \times \omega \tag{7.18}$$

where:
C_S – constant

By replacing all its components in equation (7.12), it results the detailed expression of the total retarding torque τ_r, as follows:

$$\tau_r = 0{,}5 \times C_1 \times C_D(Re) \times A_{RE} \times \rho \times \omega^2 + C_{B1} \times \rho \times v \times \omega + C_{B2} \times R_b \times \left[G_{RE} + \frac{Q_m^2}{2 \times \rho \times A_{inlet}} \right] + C_{B3} \times \omega^2 + C_s \times \omega \tag{7.19}$$

By replacing the detailed expression of τ_r in the equation (7.11), it results the detailed equation of K factor of the reaction flowmeters with moving parts.

7.3 Configuration and Operation

Figure 7.2(a,b,c) presents the configuration of the reaction flowmeter with moving parts and with vertical collinear connections.

The measured fluid enters the flowmeter through inlet connection 1 that is rigid and sealed on flowmeter cover 2. Then, the fluid continues to flow towards reaction element 3 by passing through the space between shaft 4, that rotates, together with the reaction element, in relation to upper bearing 5 and symmetrical flow deflector 6. Besides rotation, upper bearing 5, ensures also, by a free coupling with reaction element 3, the free fluid transfer to it, through its bottom end, configured as extension of inlet connection 1. The shape of reaction element 3 repeats the basic configuration of reaction element 2, presented in Figure 7.1, being provided with two identical fluid discharge nozzles 7 and 8, placed diametrically opposite.

Nozzles 7 and 8 have an identical geometry, and each of them ensures the discharge of half of flow rate, respectively $Q_m/2$.

So configured, reaction element 3 ensures the detection of the driving (reaction) torque τ_d of the two opposite and equal reaction forces F_R, to its central vertical rotation axis, allowing its horizontal free rotation together its central shaft 4.

This constructive configuration fully achieves the operation principle of the reaction measurement system of the reaction flowmeters with moving parts, presented in Figure 7.2.

Shaft 4 is rigidly fastened to reaction element 3 by screw 11 and nut 12.

At its bottom end, shaft 4 is guided and supported by bottom bearings 13 and 14, the last being fixed in three equidistant points by the pairs of screw 16 and nut 17, on the plate 15, shaped (Figure 7.2b) in order to ensure both a minimum disturbance of the fluid discharge, and its rigid fixation on circular support 20.

In its turn, support 20 is reinforced by welding it to cover 2.

This rigid and firm fixation of the both supports of upper bearing 5 and bottom bearings 13 and 14 on a common fixed support (respectively cover 2), ensures a rigorous maintenance of shaft 4 verticality during its rotation together with reaction element 3, and so is achieved implicitly a rigorous measurement of the rotation frequency, f, of reaction element 3, respectively an accurate measurement of the mass flow rate Q_m.

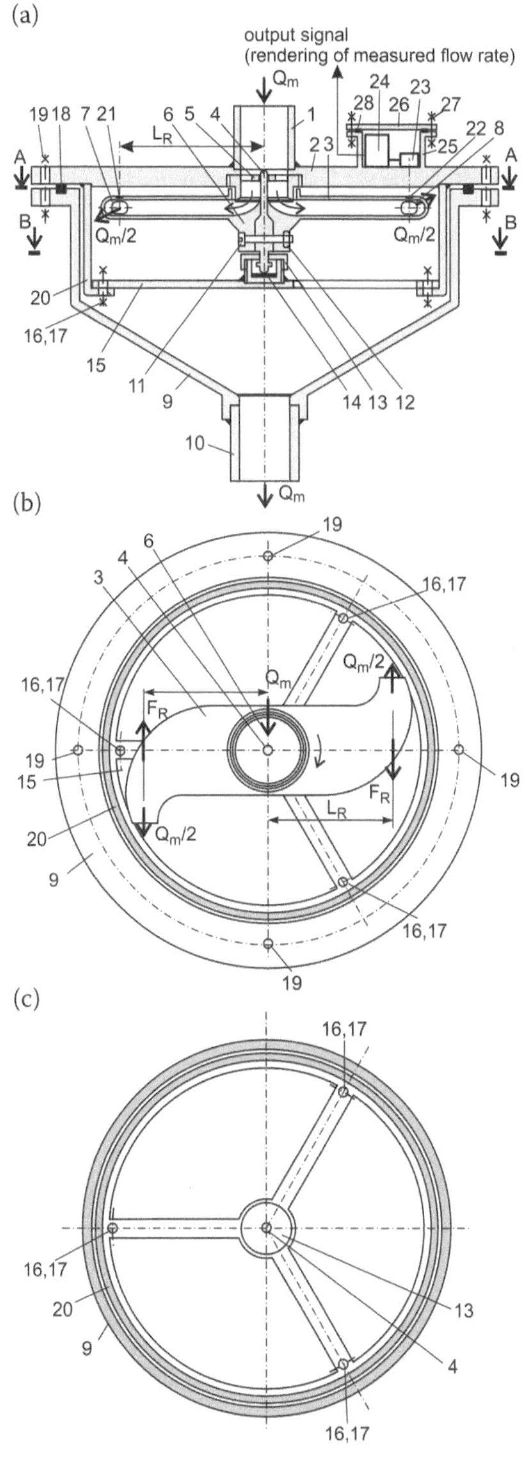
(a)
output signal
(rendering of measured flow rate)
Q_m
L_R
$Q_m/2$
$Q_m/2$
A
B
1 2 3 4 5 6 7 8 9 10 11 12 13 14 15 16,17 18 19 20 21 22 23 24 25 26 27 28
(b)
Q_m
F_R
L_R
$Q_m/2$
3 4 6 9 15 16,17 19 20
(c)
4 9 13 16,17 20

FIGURE 7.2
Reaction flowmeter with moving parts and vertical collinear connections a – transversal section b – horizontal section with a plane A – A c – horizontal section with a plane B – B
Legend: 1 – immobile inlet connection, 2 – flowmeter cover, 3 – reaction element, 4 – shaft, 5 – upper bearing, 6 – flow deflector, 7,8 – discharge nozzle, 9 – vat, 10 – outlet connection, 11 – screw, 12 – nut, 13,14 – bottom bearing, 15 – plate, 16,17 – nut/screw system, 18 – gasket, 19 – nut/screw system, 20 – support, 21,22 – electromagnetic films, 23 – magnetic pickoff sensor, 24 – electronic block, 25 – housing of outer box, 26 – cover of outer box, 27 – nut/screw system, 28 – gasket.

The fluid coming out of the reaction element in vat 9, shaped as a converging tube, is discharged from the flowmeter through outlet connection 10 that is rigid and sealed on it. Connections 1 and 10 are vertical collinear.

Cover 2 is sealed to vat 9 by tightening gasket 18, by means of nut/screw system 19. At the top of reaction element 3, into its wall two ferromagnetic small films 21 and 22, are symmetrically inserted. On cover 2, on the same radius from the RE rotation axis with the ferromagnetic films 21 and 22, is placed magnetic pickoff sensor 23 (similarly can be used other proximity sensor) that detects and measures the rotation frequency of RE, and transfers this information to electronic block 24.

Magnetic pickoff sensor 23 and electronic block 24 are placed inside an outer box, which has its own housing 25, which is closed by cover 26 with nut/screw system 27, which tightens gasket 28.

Electronic block 24 calculates the measured flowrate Q_m, by processing the measured frequency, f, according to the previous stored flowmeter calibration curve, and renders it to its exit.

7.4 Experimental Results. Analysis

Further on are presented the results of the experimental calibration, successively achieved with water and air, for the reaction flowmeters with moving parts, related to Figures 7.1 and 7.2. For their experimental calibration is used a version of flowmeter factor, $K_f = f/Q_m$ [kg^{-1}], based on the frequency output, f, by the passage of each RE arm. The frequency expression is $f = \omega \times n/(2\pi)$, where n is number of RE arms (respectively n = 2, according to its configuration presented in Figures 7.1 and 7.2). Accordingly, between K_f and the flowmeter factor K [rad/s], based on angular velocity ω, is the following correlation $K_f = K \times n/(2\pi)$.

7.4.1 Calibration with Water

A reaction flowmeter of DN 32 has been calibrated with water by the gravimetric method.

TABLE 7.1

Experimental Calibration with Water of the Reaction Flowmeter with Moving Parts

Frequency f Hz	Average Measured Flow Rate Q_m kg/h	Flowmeter Factor K_f kg^{-1}	Pressure Drop Δp bar
2,6836	70,2456	79,4774	0,014
3,7658	121,5560	79,6479	0,015
4,8859	170,2100	79,7688	0,018
9,9817	220,5025	79,8101	0,020
15,4051	450,2450	79,7807	0,030
19,4871	695,1350	79,6819	0,041
29,8615	880,4200	79,6168	0,053
40,9479	1350,2350	79,5871	0,064
51,0694	1852,2140	79,5800	0,089
62,0211	2310,2500	79,5875	0,110
72,6801	2805,4123	79,5974	0,129
84,8191	3287,1468	79,6056	0,155
108,1302	3835,7654	79,6162	0,181
125,8343	4889,3149	79,6212	0,242
144,9124	5689,4765	79,6252	0,299
166,7312	6551,7456	79,6272	0,370
175,9478	7538,0278	79,6326	0,470
186,9252	7954,1723	79,6347	0,511

Table 7.1 presents the results of this experimental calibration, and the measured flow rate Q_m values, corresponding to the rotation frequency, *f*, of the flowmeter reaction element.

In connection with Table 7.1, in Figure 7.3 is plotted the dependence curve of "K_f flowmeter factor" by Q_m. The water density was practically constant during the calibration, at the temperature t = (22 ± 0,2)°C.

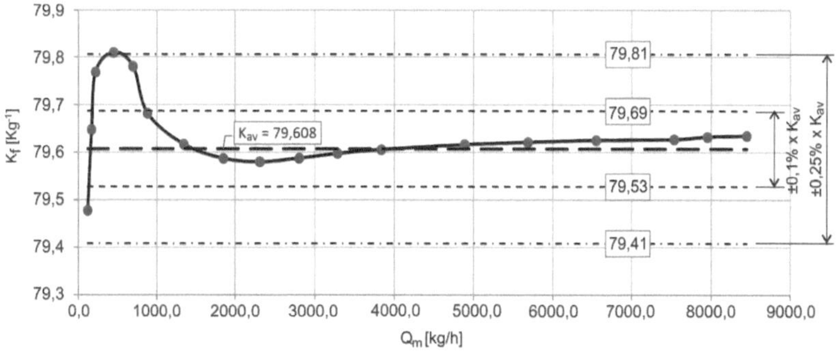

FIGURE 7.3
K_f flowmeter factor depending on water mass flow rate, for the reaction flowmeters with moving parts.

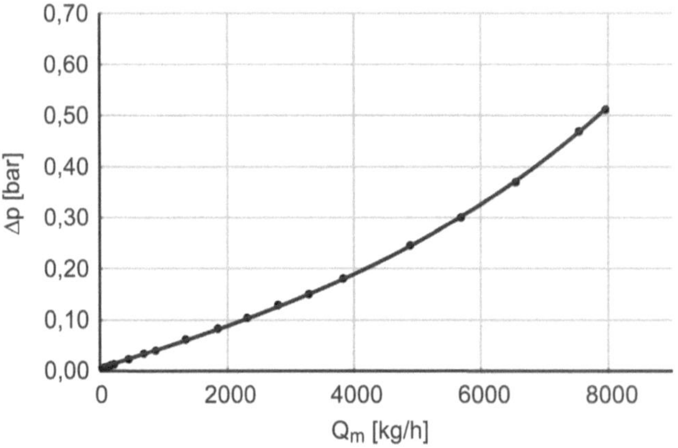

FIGURE 7.4
Pressure drop depending on measured water flow rate.

Analyzing the whole RE rotation frequency range of (2,6836 ... 186,9252) Hz it is observed that, for a linearity of ±0,25% o.r. corresponds a measured flow range of (450,1151 ... 7954,1723) kg/h and a turndown of 17,6714, respectively, for a better linearity of ±0,1% o.r. corresponds a smaller flow rate range of (880,2271 ... 7954,1723) kg/h, and a turndown of 9,0365.

The pressure drop Δp, depending on the flow rate, is presented in Figure 7.4.

7.4.2 Calibration with Air

The master slave method was used for the calibration of the reaction flowmeter.

Table 7.2 presents the results of the experimental calibration with air of the reaction flowmeter.

In connection with Table 7.2, Figure 7.5 presents the K_f flowmeter factor, depending on the air mass flow rate Q_m for the reaction flowmeters with moving parts.

Corresponding to the measured RE rotation frequency range of (14,2500 ... 1253,9580) Hz was achieved, for a linearity of ±0,5% o.r.,. the measured flow rate range of (3,1989 ... 65,9804) kg/h, and the turndown of 20,6259 and respectively for a better linearity of ±0,25% o.r., a smaller flow rate range of (5,0121 ... 65,9804) kg/h, and a turndown of 13,1642.

The pressure drop Δp depending on the air flow rate is presented in Figure 7.6.

The analysis of the previously presented results, regarding the first experimental testing of the reaction flow meters with moving parts, indicates some preliminary conclusions:

TABLE 7.2

Experimental Calibration with Air of the Reaction Flowmeter with Moving Parts

Frequency	Input Parameters				Average Measured Flow Rate	Flowmeter Factor	Pressure Drop
	Pressure	Temperature	Relative Humidity	Density			
f Hz	P bara	t °C	rH %	ρ kg/m^3	Q_m kg/h	K_f kg^{-1}	Δp bar
14,2500	0,95532	21,420	40,7900	1,1212	0,7590	67.588,9328	0,0002
14,6270	0,94320	23,671	42,6900	1,1065	0,7702	68.368,2160	0,0002
17,1740	0,94550	23,681	43,0900	1,1110	0,9001	68.688,3680	0,0004
29,5990	0,95200	24,121	42,6400	1,1151	1,5500	68.746,0645	0,0004
53,4680	0,95180	25,510	37,4100	1,1051	2,8001	68.742,1164	0,0005
108,4550	0,95090	25,890	36,6900	1,1031	5,7021	68.472,6680	0,0009
190,5090	0,94680	23,960	43,4900	1,1018	10,0421	68.295,7150	0,0024
380,6640	0,95680	25,310	39,5000	1,1049	20,0751	68.263,1917	0,0087
569,7260	0,97000	27,580	35,0600	1,1087	30,0425	68.270,4036	0,0198
666,2880	0,98730	29,120	31,8700	1,1192	35,1251	68.288,3978	0,0295
763,9020	0,94351	22,721	48,3410	1,1041	40,2541	68.317,1950	0,0398
950,2320	0,94683	21,450	40,8210	1,1210	50,0254	68.381,9660	0,0635
1.140,2880	0,95678	25,310	39,5110	1,1051	60,0025	68.414,4294	0,0993
1.253,9580	0,97021	27,580	35,0610	1,1091	65,9804	68.418,0272	0,1261

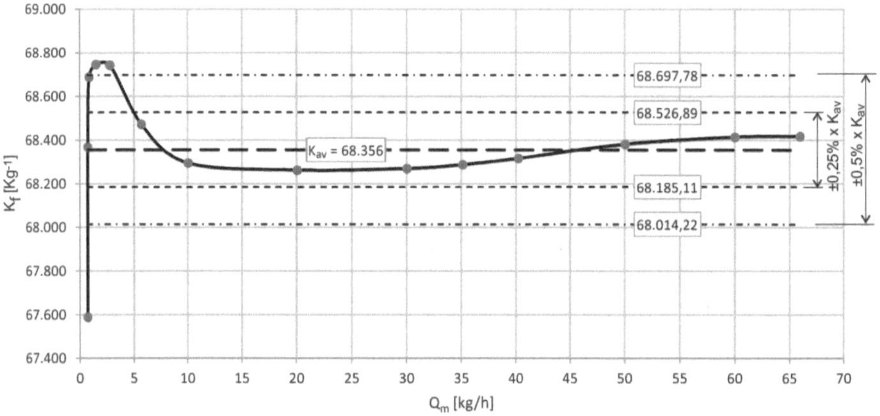

FIGURE 7.5
K_f flowmeter factor depending on the air mass flow rate of the reaction flowmeters with moving parts.

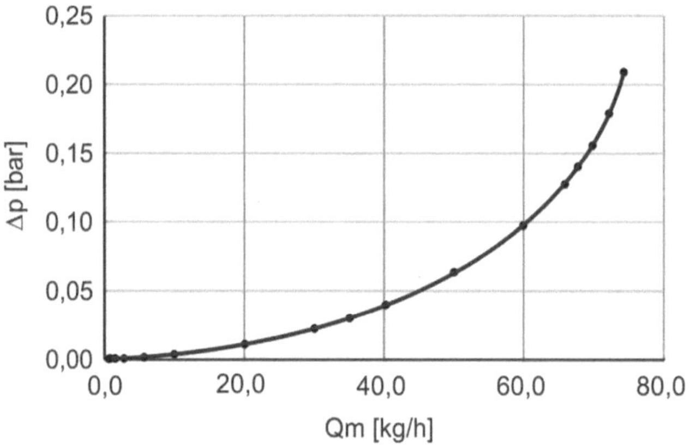

FIGURE 7.6
Pressure drop depending on the measured air flow rate.

a. a good linearity, respectively of (0,10 .... 0,25)% for measuring of water, and of (0,25 .... 0,50)% for measuring of air.
b. corresponding to these values of linearity, there are ensured reasonable values of turndown, respectively of 9 .... 17,6 for measuring of water, and of 13,1 ... 20,6 for measuring of air. The analysis of this new basic type of reaction flowmeters will be continued with the testing of flowmeters with smaller and greater nominal diameters than DN25, both for water and air and for liquids with higher viscosity.

8

Bypass Type Reaction Flowmeters

8.1 Configuration and Operation

A variant of the above presented reaction flowmeters is named "bypass type reaction flowmeters".

According to their name, the configurations of these reaction flowmeters are achieved, following the general known scheme of the bypass flowmeters, by placing on the main pipe (with high values of diameters) a pressure drop element and, in parallel with it, a pipe with a small diameter, on which is located the reaction flowmeter.

Industrial use of these bypass type reaction flowmeters is advantageous for the measurement of high fluid flow rates, related to conduits with large diameters.

Thus, the high values of the flow rates (related to the conduits with large diameters) are measured, not by using large reaction flowmeters, corresponding to them, but small reaction flowmeters, for the measurement of the low fluid flow rates, taken by the pass conduit having a small diameter.

In this respect, the configurations of these reaction flowmeters are presented in Figure 8.1.

The measuring fluid enters through inlet connection 1, corresponding to the main conduit, which has the large diameter D and is provided with a pressure drop Δp element, achieved in three variants: orifice plate 2a, nozzle 2b, or Venturi tube 2c. Bypass conduit 3 with a small diameter D1 is connected with the main conduit 1, and so a small amount q_m of the total flow rate Q_m is diverted and sampled. This small flow rate q_m is measured by the reaction flowmeter 4.

Then, the entire flow rate Q_m is discharged through the outlet connection 5 of the bypass type reaction flowmeter.

8.2 Functional Equation

The relationship between the main flow rate Q_m and the secondary flow rate q_m is ensured by the correlation between the fluidic resistance of main

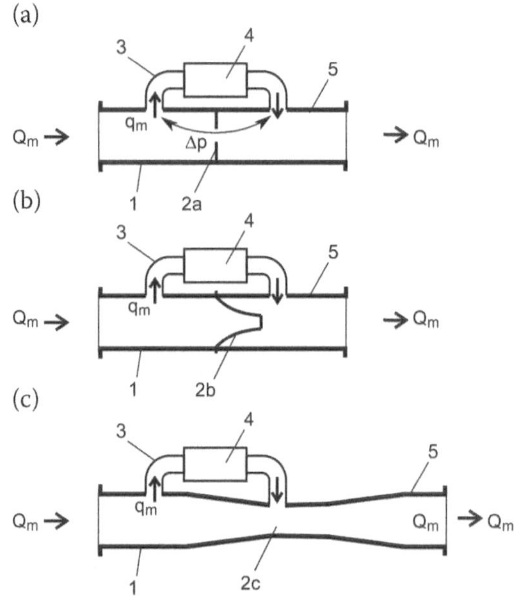

FIGURE 8.1
Configurations of the bypass type reaction flowmeters a. bypass type reaction flowmeter with orifice plate b. bypass type reaction flowmeter with nozzle c. bypass type reaction flowmeter with Venturi tube
Legend: 1 – inlet connection (main conduit), 2a – orifice plate, 2b – nozzle, 2c – Venturi tube, 3 – by pass conduit, 4 – reaction flowmeter, 5 – outlet connection

conduit 1 (of the pressure drop element) and of the bypass line (of the reaction flowmeter 4).

On the one hand, with reference to fluid flow bypass conduit 3, it results in the following dependence between the secondary (bypass) flow rate q_m and the pressure drop Δp on the fluidic restriction 2(a, b, c):

$$q_m^2 \times (R_{BP} + R_{RF}) = \Delta p \tag{8.1}$$

where:
R_{BP} – fluidic resistance of the bypass conduit (branch)
R_{RF} – fluidic resistance provided by reaction flowmeter 4

On the other hand, with reference to main conduit 1, the pressure drop Δp has the expression:

$$\Delta p = (Q_m - q_m)^2 \times R_M \tag{8.2}$$

where:
R_M – fluidic resistance provided on main conduit by fluidic restriction 2a (respectively 2b or 2c)

Replacing Δp from equation (8.2) in equation (8.1) results in the following relationship:

$$q_m^2 \times (R_{BP} + R_{RF}) = (Q_m - q_m)^2 \times R_M$$

By processing, it results in the following functional equation:

$$Q_m = q_m \times \left[1 + (R_{BP} + R_{RF})^{\frac{1}{2}} \times R_M^{-\frac{1}{2}}\right] \tag{8.3}$$

The above equation expresses the analytical dependence between the total flow rate Q_m, to be measured, entered in main conduit 1, and the secondary (bypass) flow rate q_m, effective measured by the reaction flowmeter 4.

The fluidic restriction can be achieved according to the constructive solutions generally usable for bypass type reaction flowmeters: orifice plate, nozzle, and Venturi tube. Graphical rendering of this dependency between Q_m and q_m is nonlinear in its initial portion, corresponding to the small values of the Q_m flow rate.

Then, the main portion of this graphical rendering of the dependence $Q = Q(q_m)$ is characterized by an approximately linear dependence between the two flow rates Q_m and q_m. After it, follows a new nonlinearity of the dependence between Q_m and q_m.

The linear portion of the dependence curve between the two rates Q_m and q_m is preferable to use.

Regarding the field of application of the bypass type reaction flowmeters, it is important to mention that this type of reaction flowmeter is beneficial for conduits with large diameters of DN ≥ 75 mm, for the measurement of flow rates both of liquids and gases, with a smaller upper limit of DN for gases.

The corresponding reaction flowmeter placed on the bypass conduit has a much smaller nominal diameter of (10 … 25) mm.

The diameter of the bypass conduit should be established so that the fluid velocity through it will be high enough to avoid sediment deposits.

Regarding the achievement of the pressure drop Δp element, the first variant (2a) of the orifice plate is cheaper, with an easier installation but a higher unrecovered value of Δp. The second variant (2b) has a lower recovered pressure drop and ensures the measurement of higher flow rates for the same nominal diameter of the main conduit, but it has a higher cost and more difficult mounting. The third variant (2c) has the lowest value of the unrecovered pressure drop Δp and the highest cost.

In conclusion, bypass type reaction flowmeters are advantageous for measurement, with a smaller accuracy than the individual type reaction flowmeters, for high fluid flow rates (corresponding to conduits with large diameters); they are a cheaper and less bulky solution than using an individual type of flowmeter.

Part III

Extended Reaction Flowmeters

9

Extended Reaction Flowmeters

9.1 Target of Elaboration "the Extended Reaction Force Method of Flow Measurement"

The binomial of correlated characteristics (turndown-accuracy) renders the essential technical characterization of any type of flowmeter, implicitly of the reaction flowmeters.

This consideration led us to focus on achieving a flow measurement method and the corresponding flowmeters configuration meant to ensure maximization of the turndown value, correlated with an optimal range of the accuracy values.

In this respect, a remarkable increase of the turndown (correlated with an optimal range of the accuracy values) was obtained for the reaction flowmeter by designing and achieving a new flow measurement method that uses an original type of basic configuration constituted by the coupling of two functionally interrelated reaction flowmeters.

This configuration it is generically called the extended reaction flowmeter (ERF).

The ERF uses the reaction flowmeters without moving parts, previously presented, interconnected in a specific configuration, dedicated to this purpose, according to a basic configuration scheme.

All ERFs have the same basic configuration scheme, which can have different variants of embodiment.

9.2 Principle of "the Extended Flow Measurement Method"

The Extended Reaction Force Method of flow measurement (abbreviated "ERF method") has been imagined, pursuing the achievement of the purpose presented above, starting from the specificity of the configuration and from the operation of RF without moving parts, whose configuration and operation was elaborated by the author and presented in Chapters 5.4 and 5.5.

On these bases, the method provides first of all, the need to achieve a special interconnection configuration of two such reaction flowmeters, for which should be ensured a special correlation between the values of their ranges.

Secondly, the method establishes the algorithm of functional correlation of the ranges of the two component reaction flowmeters, necessary to be met by their interrelated operation.

Thus, the ERF method indicates the basic requirements necessary to be met by the configuration of each component reaction flowmeter and the configuration of their overall interconnection, the ERF.

In this respect, the ranges of the connected component flowmeters must be consecutive and complementary, having the values of one continuing with those of the other, with "a very small common overlap zone" between them.

Thus, the entire ERF range is constituted by the summing of the range of the small flow rate values, named "the secondary range" (respectively $Q_{S_{min}} \ldots Q_{S_{max}}$) of its corresponding flowmeter, named "the secondary flowmeter", with the "the main range" of the higher values of the flow rate (respectively $Q_{M_{min}} \ldots Q_{M_{max}}$) of its corresponding flowmeter, named "the main flowmeter", minus their common overlap zone. The common overlap zone, noted ΔQ_m, has the expression:

$$\Delta Q_m = Q_{S_{max}} - Q_{M_{min}} \tag{9.1}$$

This very small common overlap zone covers both the upper end of the secondary range and the lower end of the main range, and implicitly the entire flow rate range in which the flow measurement is switched from one flowmeter to another and vice versa.

Thus, the provision of this common overlap zone of the two ranges ensures the complete continuity of the flow measurement during and along its functional switching from one flowmeter to another and vice versa.

This method provides by its operating algorithm the switching of the flow measurement from one flowmeter to another, only when the value of the instantaneous flow rate is in the immediate vicinity of the critical border of the overlap zone (lower or upper), corresponding to the direction of increase or decrease of the measured flow rate.

The detailed explication of the operation according to the algorithm of this method will be presented in Chapter 9.3.

It is necessary to mention that the share of the common overlap zone of the two ranges, referring to the whole range of the ERF, is minimal, or 1–2,5% of $Q_{M_{max}}$.

In conclusion, the ERF method achieves the proposed target by maximizing the turndown value of the ERF, correlated with an optimal range of accuracy expressed as percentage of reading of rate, respectively, E_{ERF} (% o.r.).

9.3 General Characterization of the Extended Reaction Flowmeters (ERF)

As already mentioned, practical application of the above ERF method led to the realization of a new configuration of reaction flowmeters, named ERFs.

Next is a general characterization of ERFs, including their basic configuration, correlated with their specific operation, their functional equations, and finally two practical configurations of them, elaborated for the first time in the world.

Then, there is an analysis of some case examples, so logically correlated as to sweep a very wide values range of the ERF turndown, in permanent correlation with the ranges of their corresponding accuracy.

9.3.1 Basic Structural Configuration and Operation

Starting from the previsions of the above presented method, we elaborate on the basic configuration of ERFs, shown in Figure 9.1.

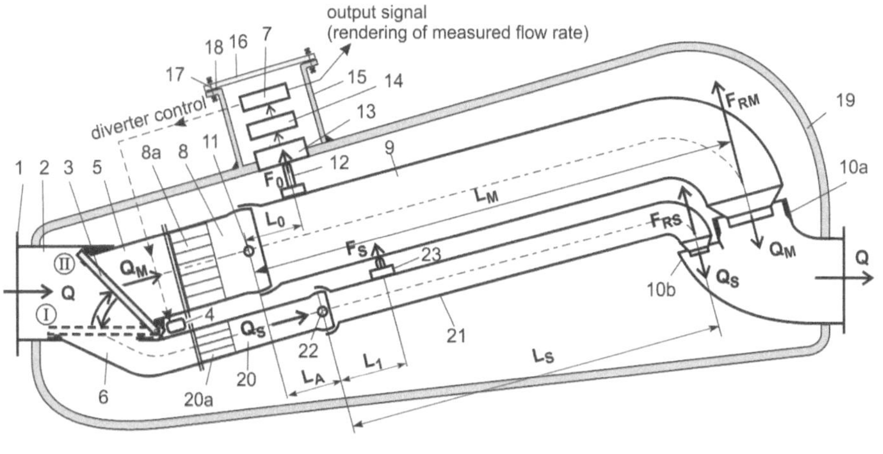

FIGURE 9.1
Basic structural scheme of extended reaction flowmeters (ERFs).
Legend: 1 – inlet connection of ERF, 2 – inlet connection of solenoid diverter, 3 – jet diverter, 4 – solenoid, 5,6 – outlet connections of solenoid diverter, 7 – electronic block, 8 – immobile inlet tube of main flowmeter, 8a – flow conditioner, 9 – reaction tube of main flowmeter, 10a,10b – joined discharge connections of "main flowmeter", respectively of "secondary flowmeter", 11 – shaft of main reaction tube, 12 – pin of main reaction tube, 13 – pushing force pick-up system, 14 – force sensor, 15 – housing of ERF outer box, 16 – cover of ERF outer box, 17 – nut/screw system, 18 – gasket, 19 – ERF housing (ERF body), 20 – immobile inlet tube of secondary flowmeter, 20a – flow conditioner, 21 – reaction tube of secondary flowmeter, 22 – shaft of secondary reaction tube, 23 – pin of secondary reaction tube.

This basic overall configuration is common for all variants of the ERF embodiments, all of them having the horizontal collinear inlet and outlet connections.

According to the ERF method, presented above, the ERF configuration ensures the functional and, implicitly, constructive correlation between its two component reaction flowmeters.

In this respect the ERF consists of a specific coupling of a "main flowmeter" with a "secondary flowmeter", having their flow ranges with correlated values, according to the previsions of this method.

As shown in Figure 9.1, the measured flow rate Q_m enters the ERF through inlet connection 1, constituted practically by inlet connection 2 of solenoid jet diverter 3, placed at the ERF inlet.

Diverter 3, acted by its solenoid 4, ensures the switching of the entered fluid to either outlet 5, which is connected with the main flowmeter (corresponding to diverter position I), or outlet 6, which is connected with the secondary flowmeter (corresponding to diverter position II), and vice versa, according to the command received from electronic block 7.

Both component flowmeters of ERF are positioned in the horizontal plane.

"The main flowmeter" is made up by its measurement system (the pair of immobile inlet tube 8 and reaction tube 9) and fluid discharge tube 10a.

Immobile inlet tube 8 can be provided with flow conditioner 8a.

Reaction tube 9 is specifically configured, being provided at its inlet with shaft 11, perpendicular on it. Thus, it is ensured a potential rotation mobility of tube 9 about the vertical axis of this shaft, and the facility to measure rigorously the force F_{R_M} (generated by the flow rate Q_M) by the accurate measurement of its rotation effect, respectively the pushing force F_0.

Reaction force acts along the axis of the tube elbow at a distance L_M from the axis of shaft 11.

On reaction tube 9, at a distance L_0 from the axis of shaft 11, is placed pin 12, which acts the pushing force pick-up system 13. In turn, the system 13 acts the force sensor 14 with the pushing force F_0, proportional to the reaction force F_{R_M}.

Thus, the electrical output signal of sensor 14 is proportional to the value of the measured pushing force F_0 (respectively to reaction force F_{R_M}) and implicitly to the measured flow rate Q_M.

The output signal of sensor 14 is applied to the input of the electric block 7, which has two simultaneous functions:

- calculation the values of the instantaneous mass flow rate Q_M
- control of the diverter position, according to Q_M values following an operating algorithm.

The output pushing force pick-up system 13, together force sensor 14 and electronic block 7, are properly positioned inside an outer box of ERF, which has its own housing 15, which is closed by the cover 16 with the nut/screw system 17, which tightens gasket 18.

Housing 15 evidently is not in contact with the measured fluid, and is rigidly fixed to ERF housing 19, which is permanently in contact with the fluid.

Force sensor 14 is rigidly fixed to housing 15 in a properly positioning in relation to the pushing force pick-up system, to be ensured the rigorous measurement of the output pushing force F_0.

Similarly, the secondary flowmeter is constituted by its measurement system (the pair of immobile inlet tube 20 and reaction tube 21) and discharge tube 10b.

Immobile inlet tube 20 can be provided with flow conditioner 20a.

Flow conditioners 8a and 20a are used optionally, when the flow process requires to reduce swirls and to remove distorted flow profiles, the flow stability being thus obtained downstream of them.

Also, reaction tube 21 is provided with vertical shaft 22, which ensures the potential rotation mobility of the tube about the shaft axis, and in this way, the facility to accurately measure the pushing force, representing the effect of the reaction force F_{R_S} generated by the flow rate Q_S. The reaction force F_{R_S} acts along the axis of the elbow tube, respectively at a distance L_S from the axis of shaft 21.

On reaction tube 21, at a distance L_1 from shaft 11, is placed pin 23, which pushes reaction tube 9 of the "main flowmeter" with the force F_S. Under the effect of this pressing force F_S, in turn reaction tube 9 presses force sensor 14 with the force F_0, which is proportional to the force F_S and implicit with the measured flow rate Q_S.

Then, the output signal of sensor 14, being proportional with the measured flow rate Q_S, is applied to electronic block 7, which ensures the calculation of the measured flow rate Q_S, operating similarly to the previous presentation regarding "the main flowmeter".

Fluid discharge tubes 10a and 10b are joined constructively in a single item, the outlet connection of the ERF.

Inlet connection 1 and common outlet connection 10a/10b are horizontal collinear, and both are fixed tightly on ERF housing 19.

Electronic block 7 performs the calculation of the instantaneous measured mass flow rate Q_m, and depending on its values, commands the switching of the diverter position according to an algorithm, ensuring its operation according to the functional correlations presented in Table 9.1.

The complete continuity of the flow measurement along the entire ERF range is achieved by ensuring only a very small overlap zone ΔQ_m of the ranges of the two component flowmeters of ERF (the main flowmeter and the secondary flowmeter), as presented in the functional diagram in Figure 9.2.

The overlap zone ΔQ_m has a very small extension, respectively:

TABLE 9.1

Algorithm of the Correlation Between Measured Flow Rate and the Diverter Status

Value of the Measured Flow Rate Q_m	Status (Position) of the Diverter	Status of the Two Lines of the Fluid Flow	
		Main Line	Secondary Line
$Q_{S_{min}}$ ———— $Q_{M_{min}}$	Position II	Open	Closed
$Q_{S_{max}}$ ———— $Q_{M_{max}}$ where: $Q_{S_{max}} > Q_{M_{min}}$	Position I	Closed	Open
$Q_S \uparrow = Q_{M_{min}}$ (Only, at the touching of value $Q_{M_{min}}$, for the increase tendency of the Q_S flow rate value)	Switching from Position II to Position I	Transition: open to closed	Transition: closed to open
$Q_M \downarrow = Q_{S_{max}}$ (Only, at the touching of value $Q_{S_{max}}$, for the decrease tendency of the Q_M flow rate value)	Switching from Position I to Position II	Transition: closed to open	Transition: open to closed

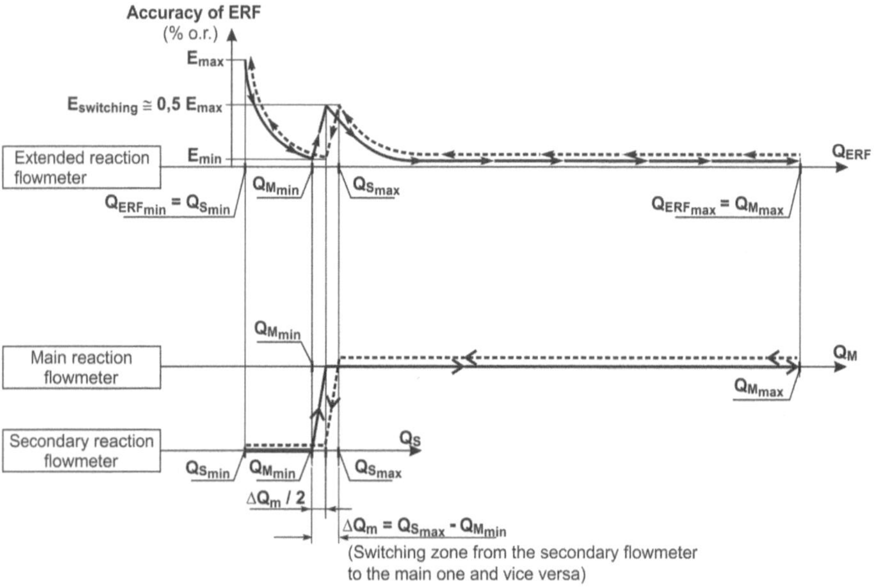

FIGURE 9.2
Functional diagram and performances of the extended reaction flowmeters.

$$\Delta Q_m = Q_{S_{max}} - Q_{M_{min}} = c \times Q_{M_{max}} \tag{9.2}$$

where:
$c = 0{,}01 \ldots 0{,}025$

Following this functional algorithm of ERF, electronic block 7 computes, on the one hand, the instantaneous value of the measured flow rate Q_m and, on the other hand, according to this value controls the position of diverter 3.

Thus is achieved the measurement of entire range ($Q_{S_{min}} \ldots Q_{M_{max}}$) of ERF, which is marked with a thickened line in Figure 9.2.

As shown in Figure 9.2, the overlap zone ΔQ_m of the range of those two component flowmeters, practically covers completely the switching inertia zone of diverter 3.

Thus are achieved both the continuity of flow measurement on entire ERF range and, along the overlap zone ΔQ_m, a moderate value of accuracy with the value of about 50% of maximum ERF accuracy value E_{max}, respectively:

$$E_{switching\ max}(\%\ \text{o.r.}) \cong 0,5 \times E_{max}(\%\ \text{o.r.}) \tag{9.3}$$

9.3.2 Functional Equations

9.3.2.1 Equation of ERF Pushing Force

This analysis is related to Figure 9.1.

Following the ERF algorithm of operation presented above, we will establish the equations of its output pushing force F_0, corresponding to the two functional positions of diverter 3.

- Diverter in Position I

(Flow measurement performed only by the main flowmeter)
Similarly with equation (5.7), the relationship between reaction force F_{R_M} and the output pushing force F_0 is the following:

$$F_0' = \frac{F_{R_M} \times L_M}{L_0} \tag{9.4}$$

where:
L_0 – moment arm, with reference to the output pushing force F_0^{I}
L_M – moment arm, with reference to the reaction force F_{R_M} of the main flowmeter
F_0' – value of the output pushing force of ERF, corresponding to position I of diverter.

- Diverter in Position II

(Flow measurement performed only by the secondary flowmeter)
With reference to Figure 9.1, it results in the following equation:

$$F_{R_S} = \frac{L_1}{L_S} \times F_S \quad (9.5)$$

Due to the pushing with force F_S of reaction tube 9 of the main flowmeter, on pin 23 of the secondary flowmeter, the output pushing force F_0 of ERF has the following expression:

$$F_0^{II} = \frac{L_1 + L_A}{L_0} \times F_S \quad (9.6)$$

where:
F_0^{II} – value of the output pushing force of ERF corresponding to the position II of diverter
F_S – output pushing force of the secondary flowmeter
L_1 – moment arm with referrence to the output pushing force F_S
L_A – distance between the axes of shaft 11 (of the main reaction tube) and of shaft 22 (of the secondary reaction tube).

By replacing the F_S expression in the F_{R_S} equation, it results in the following:

$$F_{R_S} = \frac{L_1}{L_S} \times \frac{L_0}{L_1 + L_A} \times F_0'' \quad (9.7)$$

Important remark:

Because the same sensor 14 of the output pushing force F_0 measures this force for both positions I and II of diverter 3, it is necessary to achieve the same maximum value of the output pushing force F_0, corresponding to both maximum value of the measured flow rate, respectively $Q_{M_{max}}$ (measured by the main flowmeter, corresponding to position I of the diverter) and $Q_{S_{max}}$ (measured by the secondary flowmeter, corresponding to position II of the diverter).

In this respect, it results in the mandatory requirement $F'_{0_{max}} = F''_{0\,max}$, and by its processing is obtained the expression of the necessary correlation between the maximum values of the reaction forces generated by the two component reaction flowmeters of the ERF, according to the following equation:

$$F_{R_{M_{max}}} \times \frac{L_M}{L_0} = F_{R_{S_{max}}} \times \frac{L_S}{L_1} \times \frac{L_1 + L_A}{L_0} \quad (9.8)$$

9.3.2.2 Equation of ERF Turndown

Using the general flow expression of the reaction flowmeters (3.3), we obtained successively the flow equations of ERF for the two positions of the diverter.

- Diverter in Position I:

$$Q^{I}_{ERF} = \left(F_{R_M} \times \rho \times k^{-1}_{1_M}\right)^{1/2} \tag{9.9}$$

where:
Q^{I}_{ERF} – ERF flow rate measured only by the main flowmeter, along the range $(Q_{S_{max}} \dots Q_{M_{max}})$

- Diverter in Position II:

$$Q^{II}_{ERF} = \left(F_{R_S} \times \rho \times k^{-1}_{1_S}\right)^{1/2} \tag{9.10}$$

where:
Q^{II}_{ERF} – ERF flow rate measured only by the secondary flowmeter, along the range $(Q_{S_{min}} \dots Q_{M_{min}})$.

The analytical correlation between the turndown of the reaction flowmeters and the maximum permissible value of their accuracy expressed as a percentage of reading of the rate E_F (% o.r.), and the measurement accuracy of their output force sensor, is expressed as a percentage of the full scale rate, $E_S(\%\ o.f.s.) = E_{Force\ sensor}(\%\ o.f.s.)$, is as follows:

- turndown T_M of the main flowmeter

$$T_M = \frac{Q_{M_{max}}}{Q_{M_{min}}} = \sqrt{\frac{E_{MF_{max}}(\%\ o.r.)}{E_{Force\ sensor}(\%\ o.f.s.)}} \tag{9.11}$$

- turndown T_S of the secondary flowmeter

$$T_S = \frac{Q_{S_{max}}}{Q_{S_{min}}} = \sqrt{\frac{E_{SF_{max}}(\%\ o.r.)}{E_{Force\ sensor}(\%\ o.f.s.)}} \tag{9.12}$$

Those presented above, with reference to the basic configuration and the operation of the ERF in connection with Figure 9.2, led us to the conclusion that the total flow rate range of the ERF, respectively $Q_{S_{min}} \dots Q_{M_{max}}$, results by summing up the ranges of the two component reaction flowmeters (the main and the secondary), having consecutive values, minus the minimal functional overlap between them, ΔQ_m.

From this reason the value of the full scale of the secondary flowmeter, $Q_{S_{max}}$ is correlated with the minimum measurable flow rate of the main flowmeter, according to the expression:

$$Q_{S_{max}} = Q_{M_{min}} + \Delta Q_m = \frac{Q_{M_{max}}}{T_M} + \Delta Q_m = \frac{Q_{M_{max}} + \Delta Q_m \times T_M}{T_M} \tag{9.13}$$

It results that the value of the minimum measurable flow rate by the secondary flowmeter $Q_{S_{min}}$ has the expression:

$$Q_{S_{min}} = \frac{Q_{S_{max}}}{T_S} = \frac{Q_{M_{min}} + \Delta Q_m}{T_S} = \frac{\frac{Q_{M_{max}}}{T_M} + \Delta Q_m}{T_S} = \frac{Q_{M_{max}} + \Delta Q_m \times T_M}{T_S \times T_M} \tag{9.14}$$

Thus, knowing the specific analytical correlation between $Q_{S_{min}}$ and $Q_{M_{max}}$, we obtained the equation of the global turndown of the ERF, expressed with reference to the individual turndown expressions of the main flowmeter and of the secondary flowmeter, as follows:

$$T_{ERF} = \frac{Q_{M_{max}}}{Q_{S_{min}}} = \frac{Q_{M_{max}}}{\frac{\Delta Q_m + Q_{M_{max}} \times T_M}{T_S}} = \frac{Q_{M_{max}}}{\frac{Q_{M_{max}} + \Delta Q_m \times T_M}{T_S \times T_M}} \tag{9.15}$$

respectively:

$$T_{ERF} = \frac{T_S \times T_M}{1 + T_M \times \frac{\Delta Q_m}{Q_{M_{max}}}} \tag{9.16}$$

By replacing the values of T_M and T_S with their expressions, we obtained the detailed equation of ERFs:

$$T_{ERF} = \frac{\left[\frac{E_{SF_{max}}(\%\ o.r.)}{E_{Force\ Sensor}(\%)} \times \frac{E_{MF_{max}}(\%\ o.r.)}{E_{Force\ Sensor}(\%\ o.f.s.)}\right]^{1/2}}{1 + \frac{\Delta Q_m}{Q_{M_{max}}} \times \left[\frac{E_{MF_{max}}(\%\ o.r.)}{E_{Force\ Sensor}(\%\ o.f.s.)}\right]^{1/2}} \tag{9.17}$$

Because for both component flowmeters of the ERF ("the main" and "the secondary"), on the one hand, it uses the same force sensor and, on the other hand, it requires the same maximum value for the relative flow measuring "accuracy expressed as percentage of reading of rate", the result is as follows:

$$T_{ERF} = \frac{\frac{E_{ERFmax}(\%\ o.r.)}{E_{Force\ Sensor}(\%\ o.f.s.)}}{1 + \frac{\Delta Q_m}{Q_{Mmax}} \times \left[\frac{E_{ERFmax}(\%\ o.r.)}{E_{Force\ Sensor}(\%\ o.f.s.)}\right]^{1/2}} \qquad (9.18)$$

where:

$E_{ERF_{max}}(\%\ o.r.)$ – relative accuracy of ERF expressed as percentage of reading of rate

$$E_{ERF_{max}}(\%\ o.r.) = E_{SF_{max}}(\%\ o.r.) = E_{MF_{max}}(\%\ o.r.)$$

9.4 ERF with Pushing Force Measurement

9.4.1 ERF with Direct Measurement of Pushing Force

Structural configuration of the ERFs with direct measurement of the output pushing force is presented in Figure 9.3.

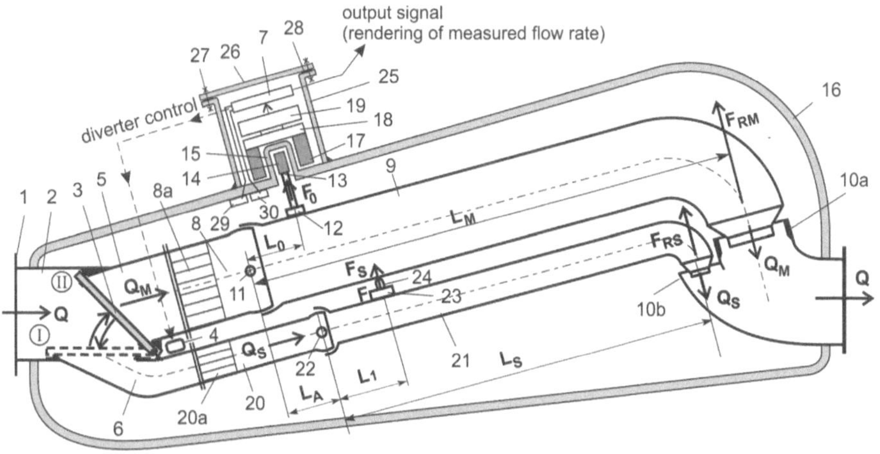

FIGURE 9.3
Structural configuration of the extended reaction flowmeters with direct measurement of output pushing force.
Legend: 1 – inlet connection of ERF, 2 – inlet connection of solenoid diverter, 3 – diverter, 4 – solenoid, 5,6 – outlet connections of solenoid diverter, 7 – electronic block, 8 – immobile inlet tube of main flowmeter, 8a – flow conditioner, 9 – reaction tube of main flowmeter, 10a,10b – joined discharge connections of "main flowmeter" respectively "secondary flowmeter", 11 – shaft of main reaction tube, 12 – boss, 13 – pin of main reaction tube, 14 – inner magnet, 15 – small closed tube, 16 – ERF housing, 17 – outer ring magnet, 18 – support, 19 – force sensor, 20 – immobile inlet tube of secondary flowmeter, 20a – flow conditioner, 21 – reaction tube of secondary flowmeter, 22 – shaft of secondary reaction tube, 23 – boss, 24 – pin of secondary reaction tube, 25 – housing of ERF outer box, 26 – cover of ERF outer box, 27 – nut/screw system, 28 – gasket, 29 and 30 – sensors of P and T.

As shown in Figure 9.3, the measured flow rate Q enters the ERF through its inlet connection 1, consisting of inlet connection 2 of the solenoid diverter, placed at the ERF inlet.

Diverter 3, being acted by its solenoid 4, switches the flow of the entered fluid to either outlet 5, to the main flowmeter (corresponding to the position I of the diverter), or outlet 6, to the secondary flowmeter (corresponding to the position II of the diverter) and vice versa, according to the command received from electronic block 7.

Let us consider that diverter 3 is initially commanded in position I, and consequently *the incoming fluid is directed to flow only through the main flowmeter.*

The main flowmeter consists of immobile inlet tube 8, reaction tube 9 that generates the reaction force F_{R_M}, and its discharge tube 10a.

The main reaction tube 9 is provided with shaft 11, perpendicular on it and rigid with it, which ensures a potential rotation mobility of the tube around the vertical axis of this shaft.

The detailed explanation of the constructive configuration, which ensures the rigorous positioning of shaft 11, was presented in Chapter 5.5.

The reaction force F_{R_M}, generated by reaction tube 9, acts along the axis of elbow tube, at a distance L_M from the shaft axis.

The main reaction tube 9 is provided with boss 12. In this boss, at a distance L_0 from the shaft axis, pin 13 is rigidly embedded. On the end of pin 13 is rigidly fixed magnet 14, which is placed inside small closed tube 15.

The couple of pin 13 and magnet 14 can move freely in relation to tube 15, which is configured as a part of ERF housing 16.

Outside tube 15, around it, is placed ring magnet 17, with a free displacement related to it, its inner hole having a suitable greater diameter than the outer diameter of tube 15.

Magnet 17 is rigidly fixed to its support 18, which, in turn, is constructively fixed to the mobile part of force sensor 19, and permanently transmits it the pushing force F_0, received by magnetic coupling from magnet 14.

The value of the output pushing force F_0 is proportional to the reaction force F_{R_M}, according to those presented above in Chapters 5.2.3, 5.5, and 9.3.2.

Force sensor 19 receives the output pushing force F_0 (with the value F_0^{I}, corresponding to position I of diverter 3), and following its principle of operation, to any very small tendency of displacement of its mobile part (corresponding to any small variation of the force F_0^{I}) it achieves its electromagnetic balancing and its measurement, in the same time.

Consequently, the electrical output signal of force sensor 19 is proportional to the value of the pushing force F_0^{I} and is applied to electronic block 7.

By continuing the presentation, based on Figure 9.3, ring magnet 17, its support 18, force sensor 19, and electronic block 7 are properly positioned inside the outer box of ERF, which has its own housing 25, which is closed by cover 26 with the nut/screw 27, which tightens gasket 28.

Housing 25 is not in contact with the measured fluid and is rigidly fixed to housing 16 (the body flowmeter), which is permanently in contact with the fluid.

Force sensor 19 is rigidly fixed to housing 25, in a proper position in relation to the pushing force pick-up system, in order to ensure a rigorous measurement of the pushing force.

Electronic block 7 calculates the corresponding value of the reaction force F_{R_M}, according to the known relationship between forces F_0^{I} and F_{R_M}, and then the corresponding instantaneous value of the measured main flow rate Q_M, using the calibration curve $Q_M = Q_M\ (F_{R_M})$, previously stored in its memory. The flow rate Q_M is compensated with pressure P and temperature T, by sensors 29 and 30.

This is the first of the two functions of block 7, because as presented in Chapter 9.3.1, block 7 has two simultaneous functions (the calculation of instantaneous flow rate, and the control of diverter 3 position).

In this respect, according to the algorithm presented in Table 9.1, block 7 will maintain diverter 3 on position I, only when the values of the measured flow rate by the main flowmeter fall within the range $Q_{M_{max}} \dots Q_{S_{max}}$.

When the value of the measured flow rate by the "main flowmeter" decreases and tends to reach the value $Q_M = Q_{S_{max}}$, electronic block 7 commands the switching of diverter 3 from its initial position I to position II, and now *the incoming fluid is directed to flow only through the secondary flowmeter,* the flow measurement being continued only by it.

The secondary flowmeter consists of inlet immobile tube 20, reaction tube 21, and discharge tube 10b.

Fluid discharge tubes 10a and 10b are constructively joined in a single item, the outlet connection of ERF.

Secondary reaction tube 21 is provided with vertical shaft 22, which ensures the potential rotation mobility of the tube around the shaft axis.

Shaft 22 of secondary reaction tube 21 is functionally positioned to the right of shaft 11 of the main reaction tube 9, at a distance L_A.

The reaction force F_{R_S} generated by reaction tube 21 acts along the axis of the elbow tube, respectively at the distance L_S from its shaft axis.

On secondary reaction tube 21 is placed boss 23. In this boss, at the distance L_1 from shaft axis, a pin 24 is rigidly embedded.

The relative positioning between the main reaction tube 9 and secondary reaction tube 21 is achieved constructively so that pin 24 can be in permanent contact with the main reaction tube 9 and push it with the force F_S, throughout the period when diverter 3 is in position II, and the flow measurement is achieved by the secondary flow meter.

In this way, under the effect of this force, pin 13 of the main reaction tube 9 will press with a force, having the amplified value F_0^{II}, according to the previsions of Chapter 9.3.2.1.

The pushing force F_0^{II} is proportional to force F_S and implicitly with the measured flow rate Q_S.

Then, the operation is similar to the one presented above, corresponding to the situation when diverter 3 is in position I.

Thus, pushing force F_0^{II} is measured by sensor 19, its value being proportional to the reaction force F_S, implicitly with the measured flow rate Q_S.

Functionally, according to previsions indicated in Chapter 9.3.2, it is necessary to meet the requirement $F_{0_{max}}^{I} = F_{0_{max}}^{II}$.

The output signal of sensor 19 is applied to block 7, which, on the one hand, calculates the value of the instantaneous flow rate Q_S (compensated with P and T), and, on the other hand, commands the position of diverter 3 according to the algorithm presented in Table 9.1:

- if $Q_{S_{min}} \leq Q_S < Q_{M_{min}}$ the diverter is held on the same position II, and the flow measurement is continued by the secondary flowmeter
- if Q_S increases and tends to reach the value, the diverter is switched to return from position II to position I, and thus the flow measurement is continued by "the main flowmeter".

It is necessary to mention that, because the overlap zone of the ranges of the two flowmeters (the main and the secondary) is very small, of $\Delta Q_m = Q_{S_{max}} - Q_{M_{min}} = (0{,}01 \ldots 0{,}025) \times Q_{M_{max}}$, the switching of diverter practically does not disturb the flow measurement.

9.4.2 ERF with Differential Measurement of Pushing Force

Structural configuration of ERFs with differential measurement of output pushing force is presented in Figure 9.4.

It is necessary to specify that the configuration of this type of flowmeter is largely similar to that of a previously presented type of flowmeter, respectively "ERF with output pushing force by its electromagnetic balancing".

In this respect, by comparing the structural configuration of the two types of ERF, shown in Figure 9.3 and Figure 9.4, it follows that the essential difference between them consists in the method to measure the output pushing force and implicitly of the corresponding system to pick up this force.

Also, we mention that the principle of the method in question, the specific corresponding configuration of the force pick-up system, and the related functional equation, have been presented previously in Chapters 5.2.2 and 5.4.1.

Consequently, only the aspects specific to the ensemble operation of this type of ERF will be presented below.

As shown in Figure 9.4, diverter 3, being acted by its solenoid 4, switches the flow of the entered fluid to either outlet 5, to the main flowmeter (corresponding to position I of diverter), or outlet 6 to the secondary flowmeter (corresponding to position II of diverter) and vice versa, according to the command received from electronic block 7.

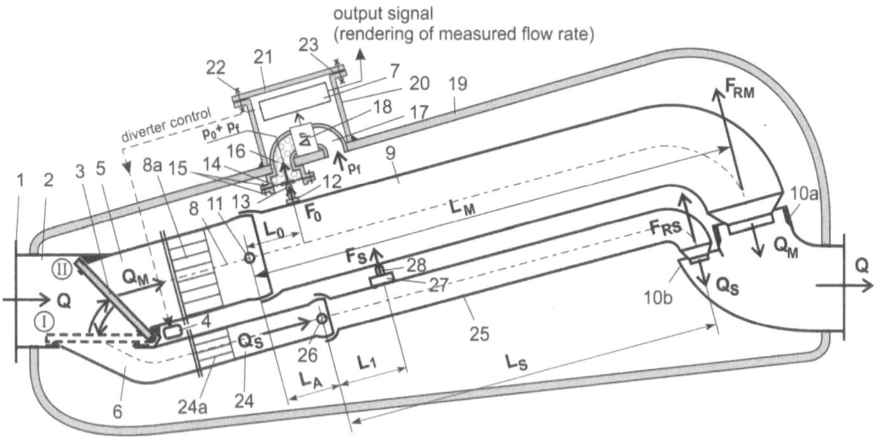

FIGURE 9.4
Structural configuration of the extended reaction flowmeters with the differential measurement of output pushing force.
Legend: 1 – inlet connection of ERF, 2 – inlet connection of solenoid diverter, 3 – diverter, 4 – solenoid, 5,6 – outlet connections of solenoid diverter, 7 – electronic block (of flow rate calculation and diverter control), 8 – immobile inlet tube of main flowmeter, 8a – flow conditioner, 9 – reaction tube of main flowmeter, 10a,10b – discharge connections of "main flowmeter" respectively "secondary flowmeter", 11 – shaft of main reaction tube, 12 – boss, 13 – pin of main reaction tube, 14 – separation membrane, 15 – flanges that tighten sealed the membrane, 16 – connection of high pressure inlet of Δp sensor, 17 – connection of low pressure inlet of Δp sensor, 18 – Δp (differential pressure) sensor, 19 – ERF housing (ERF body), 20 – housing of ERF outer box, 21 – cover of outer box, 22 – nut screw system, 23 – gasket, 24 – immobile inlet tube of secondary flowmeter, 24a – flow conditioner, 25 – reaction tube of secondary flowmeter, 26 – shaft, 27 – boss, 28 – pin of secondary flowmeter.

Let us consider that diverter 3 is initially commanded in position I and consequently *the fluid is directed to flow only through the main flowmeter.*

According to the ERF general functional algorithm, the operation of this type of ERF, respectively of its main flowmeter (corresponding to position I of diverter 3) is similar to that of the previous ERF type.

In line with the above remarks, by continuing to present the ERF type in question, compared to the previous ERF type, the first essential difference consist in the takeover way of pushing force F_0^I from pin 13.

F_0^I has the relationship as presented in Chapter 9.3.2.1.

Now, the pin presses on a face of the separation membrane 14 with force F_0^I, causing the corresponding pressure p_0 to appear on its opposite face.

The pressure p_0, to which static pressure p_f of measured fluid is permanently added, is taken up by the transmission liquid, which completely fills the connection 16, and their summation is applied to the high pressure inlet of Δp sensor, 18.

The influence of the static pressure p_f of the measured fluid, is permanently canceled by its inlet, via connection 17, to the low pressure input of Δp sensor, 18.

Thus, sensor 18 ensures, compensated with temperature, the differential measurement of pushing pressure p_O, thus obtaining a very rigorous precision of the flow rate measurement.

The output signal of sensor 18, which renders the p_O value, is applied to electronic block 7.

Electronic block 7, starting from the calculation of the corresponding value of the reaction force F_{R_M} (that determined the appearance of pressure p_O) ensures the calculation of the instantaneous value of the measured flow rate Q_M, using the calibration curve $Q_M = Q_M\ (F_{R_M})$, previously stored in its memory.

According to the ERF algorithm presented in Table 9.1, block 7 will maintain diverter 3 on position I only when the values of the flow rate Q_M, measured by the main flowmeter, fall within the range ($Q_{M_{max}} \cdots Q_{S_{max}}$).

When the value of the flow rate measured by the main flowmeter decreases and tends to reach the value $Q_M = Q_{S_{max}}$, electronic block 7 commands the switching of diverter 3 from its initial position I to position II, and now *the incoming fluid is directed to flow only through the secondary flowmeter*, the flow measurement being continued only by it.

As previously presented in Chapter 9.3.1, shaft 26 of the secondary reaction tube is functionally positioned to the right of shaft 11 of the main reaction tube, at a distance L_A.

The secondary reaction tube 25 is placed pin 28, at a distance L_1 from the axis of shift 26.

The relative positioning between the main reaction tube 9 and the secondary reaction tube 25 is constructively achieved so that pin 28 is in permanent contact with the main reaction tube 9 and pushes it with the force F_S throughout the period when diverter 3 is in position II, and the flow measurement is ensured by the secondary flowmeter.

In this way, under the effect of this force, pin 13 of the main reaction tube 9 will press the separation membrane 14 with a force having the amplified value F_0^{II}, according to those presented in Chapter 9.3.2.1.

Further the takeover and the measurement of this pushing force F_0^{II}, are achieved identically by the Δp sensor 18, and then the calculation of the measured flow rate value by electronic block 7, are achieved identically by the Δp sensor 18, as in the situation presented above, corresponding to position I of diverter 3.

According to Chapter 9.3.2, it is necessary to ensure the condition $F_{0_{max}}^{I} = F_{0_{max}}^{II}$.

By following the functional algorithm as for previous ERF type, and corresponding to the value of the measured flow rate Q_S, electronic block 7 commands the position of diverter 3, as follows:

- for $Q_{S_{min}} \leq Q_S < Q_{M_{min}}$, the diverter is held in the same position II, and the flow measurement is ensured and continued by the secondary flowmeter

- if Q_S increases and tends to reach the value $Q_S = Q_{M_{min}}$, diverter 3 switched to return from position II to position I, and thus from now on the flow measurement is continued by the main flowmeter.

As a general remark, it should be mentioned that, because the overlap zone of the ranges of the two flowmeters ("the main" and "the secondary") is very small, of $\Delta Q_m = Q_{S_{max}} - Q_{M_{min}} = (0{,}01 \ldots 0{,}025) \times Q_{M_{max}}$, the switching of diverter practically does not disturb the flow measurement.

9.5 Comparative Analysis Between ERF and its Component Reaction Flowmeters, Regarding the Binomial Dependence (Turndown - Accuracy)

Chapter 9.3.2 above presented the specific analytical characterization of the intrinsic inter-correlation between the turndown and the accuracy of ERFs.

As mentioned from the beginning, the author's target has been the imagining of a new different configuration of reaction flowmeters (named "ERF"), which ensures the maximizing of the turndown value, obtained by the reaction flowmeters types as previously presented.

Surely, the maximized turndown values need to be correlated with an optimum range of accuracy values (expressed as a percentage of reading of rate, respectively % o.r.), because this couple of features provides the essential characterizing of any type of flowmeter.

In this respect, in order to achieve a synoptic overview of the significant performance obtained by ERF, regarding these two features, the results of some relevant successive case examples, in their logical connection, are presented in Table 9.2.

This ensures, in a systematic way, the detailed analysis of the specific change for ERFs of their turndown, depending on the changing of the established values, for the reference parameters that determine it, respectively:

- $E_{F_{max}}$, the maximum permissible value of accuracy (expressed as a percentage of reading of rate, respectively % o.r.) for both component reaction flowmeters of ERF
- E_S, the accuracy (expressed as a percentage of full scale, respectively % o.f.s.) of the sensor of output pushing force, used by ERF.
- ΔQ_m, common overlap zone of ranges of the two component reaction flowmeters of ERF

Table 9.2 presents initially two case examples of ERF that, for the measuring of their output pushing force, use sensors with electromagnetic balancing of force.

TABLE 9.2

Analysis of ERF Features Depending on Their Reference Parameters

Reference Parameters of the Extended Reaction Flowmeters (ERF)				Resulting Features of ERF	
Reference Parameters of the Two Nonlinear Reaction Flowmeters, Component of ERF		Resulting Turndown of Each Nonlinear Reaction Flowmeter, Component of ERF	Overlap Zone of the Ranges of the Two Reaction Flowmeters, Component of ERF	Theoretical Turndown of ERF	Accuracy (Measurement Error) Values Range of ERF
Maximum Permissible Error of Sensor Which Measures the Output Parameter of Reaction Flowmeter	Maximum Permissible Error of Reaction Flowmeter				
E_S	$E_{F,max}$	$T_M = T_S$	$\Delta Q_m / Q_{M\,max}$	T_{ERF}	$E_{ERF} = E_M = E_S$
% o.f.s.	% o.r.	-	-	-	% o.r.
3×10^{-4}	2	81,65	0,010	3.670,09	0,0003 ... 2
			0,025	2.192,10	
	1	57,73	0,010	2.112,95	0,0003 ... 1
			0,025	1.364,06	
	0,5	40,82	0,010	1.183,27	0,0003 ... 0,5
			0,025	824,68	
	0,1	18,25	0,010	281,65	0,0003 ... 0,1
			0,025	228,71	
10^{-3}	2	44,72	0,010	1.381,98	0,001 ... 2
			0,025	944,29	
	1	31,62	0,010	759,76	0,001 ... 1
			0,025	558,51	
	0,5	22,36	0,010	408,63	0,001 ... 0,5
			0,025	320,72	
	0,1	10,00	0,010	90,91	0,001 ... 0,1
			0,025	80,00	
10^{-2}	2	14,14	0,010	175,22	0,01 ... 2
			0,025	147,77	
	1	10,00	0,010	90,91	0,01 ... 1
			0,025	80,00	
	0,5	7,07	0,010	49,65	0,01 ... 0,5
			0,025	42,49	
	0,1	3,16	0,010	9,97	0,01 ... 0,1
			0,025	9,27	
$1{,}4 \times 10^{-2}$	2	11,95	0,010	127,61	0,014 ... 2
			0,025	110,00	
	1	8,45	0,010	65,85	0,014 ... 1
			0,025	58,97	
2×10^{-2}	2	10,00	0,010	90,91	0,02 ... 2
			0,025	80,00	
$2{,}5 \times 10^{-2}$	2	8,95	0,010	73,43	0,025 ... 2
			0,025	65,37	

In this respect, the first example indicates, for ERFs that use this type of force sensor, having a high accuracy (linearity) of 0,0003% o.f.s., the following corresponding ranges were obtained for their features, respectively:

- from a theoretical maximum turndown of 3.670,09, and a corresponding accuracy values range of (0,0003 … 2)% o.r.
- up to a theoretical minimum turndown of 228,71, and a corresponding accuracy values range of (0,0003 … 0,1)% o.r.

Then, in the second case example are indicated, for ERFs that use the same type of force sensor, but with a lower accuracy (linearity), of 0,001% o.f.s., the new corresponding values obtained for the same two features of ERF, respectively:

- from a theoretical maximum turndown of 1.381,98 and a corresponding accuracy values range of (0,001 … 2)% o.r.
- up to a theoretical minimum turndown of 80 and a corresponding accuracy values range of (0,001 … 0,1)% o.r.

These two successive case examples render explicitly the trajectory followed by the turndown changing, corresponding to the decrease of the force sensor accuracy.

These examples demonstrate that, by summing the advantages determined by their specific configuration and the high accuracy of this type of force sensor, the ERF can lead to very important performance for their essential features rendered above.

After the presentation of an intermediary case example (referring to ERF using a force sensor with accuracy of E_S = 0,01% o.f.s.), the last three case examples indicate ERFs that are possible to use, with good results and other types of force sensors.

These three case examples refer to the previously presented solutions, respectively by using the load cell with strain gauge (according to Chapters 5.2.3 and 5.5.1) and the differential pressure sensor coupled with a separation diaphragm (according to Chapters 5.2.2 and 5.4).

According to these last case examples, the analyzed binomial (turndown correlated with the accuracy) of ERF has lower values,

Thus, when for the measurement of pushing force, is used a load cell with strain gauge, heaving E_S = 0.014 % o.f.s., it results that the analyzed binomial (turndown correlated with the accuracy) of ERF, varies between following theoretical limits:

From a turndown of 127,61 and a corresponding accuracy range of (0.014 … 2)% o.r., to a turndown of 58,97 and a corresponding accuracy range of (0,014 … 1) % o.r.

Also, when for the measurement of pushing force, is used a deferential pressure sensor, having E_S = 0.02 … 0.025 (% o.f.s.), coupled with a separation diaphragm, the analyzed binomial (turndown correlated with the accuracy) of ERF, varies between the following theoretical limits: from a maximum turndown of 90,91 and a corresponding accuracy values range of (0,02 … 2)% o.r., to a minimum turndown of 65,37, corresponding accuracy values range of (0,025 … 2)% o.r.

Supplementary to these special performances should be added the important advantage of the lack of contact of the force sensor with the measured fluid, and implicitly the lack of need to ensure its constructive protection against the fluid physical – chemical action.

This overview demonstrates, on the one hand, that due to their basic structural configuration the ERF can ensure very good values for the binomial (turndown correlated with accuracy), the essential features of any type of flowmeter.

On the other hand, it demonstrates their availability to be used for a very wide values of range of this analyzed binomial, with good results.

Part IV

Conclusions

10

Brief Overview on the Reaction Flowmeters Features

After the above presentation of the different basic types of the reaction flowmeters, configured by the author using his Unitary Synthesis and Design Method of Flowmeters [2.1], a brief overview on their essential operating features is considered useful.

It is known that, for any type of flowmeter, the correlation of its working range (characterized by its turndown) with its measurement accuracy, is critical to its successful applications.

For this reason, this short overview is focused in particular on the binomial given by the correlation between the measurement accuracy and the turndown of the reaction flowmeters that essentially expresses at a glance their performance.

Consequently, referring to each of the above basic types of reaction flowmeters (nonlinear, linear, extended), a comparative overview of the dependence between turndown and accuracy is achieved in two successive stages.

In this regard, the first stage is the presentation of their analytical expressions.

The second stage continues with a presentation of the results of a succession of several numerical case examples, thus correlated in order to show a global evaluation of the field of their potential applications.

10.1 Comparative Presentation of the Analytical Turndown Expressions

Further on are reiterated, for each basic type of the reaction flowmeters, the analytical relationships expressing the dependence between their turndown T and their measurement accuracy.

As previously demonstrated, for all different basic types of reaction flowmeters (nonlinear, linear, extended) the turndown relationships are expressed by the analytical correlation between the maximum permissible measurement error $E_{F/max}$ of each type of reaction flowmeter and the maximum permissible error of the sensor, that measures its characteristic parameter of the respective flowmeter.

The characteristic parameter is then applied to the input of the Secondary Element (final block) of the reaction flowmeter that, proportional to it, gives at its output the output parameter Q_R, that renders the value of the measured mass flow rate Q_m.

According to these considerations, below are presented successively the turndown expressions, established in the previous chapters, respectively:

- *For the nonlinear reaction flowmeters*

The turndown (noted T_{NRF}) has the expression (5.11), respectively:

$$T_{NRF} = \frac{Q_{m,max}}{Q_{m,min}} = \sqrt{\frac{E_{F,max}\,(\%\ o.r.)}{E_S\,(\%\ o.f.s.)}}$$

- *For the linear reaction flowmeters*

The turndown (noted T_{LRF}) has the expression (6.12), respectively:

$$T_{LRF} = \frac{Q_{m,max}}{Q_{m,min}} = \frac{E_{F,max}\,(\%\ o.r.)}{E_S\,(\%\ o.f.s.)}$$

- *For the extended reaction flowmeters*

The turndown (noted T_{ERF}) has the expression (9.19), respectively:

$$T_{ERF} = \frac{Q_{M,max}}{Q_{S,min}} = \frac{E_{F,max}\,(\%\,o.r.)}{E_S\,(\%\,o.f.s.)} \times \frac{1}{1 + \frac{\Delta Q_m}{Q_{M,max}} \times \left[\frac{E_{F,max}\,(\%\ o.r.)}{E_S\,(\%\ o.f.s.)}\right]^{1/2}}$$

According to the previous explanations, it is mentioned that on the one hand, in all these above relationships of the turndown:

- the maximum permissible measurement error (accuracy) of reaction flowmeters, is expressed as a percentage of reading, respectively $E_{F,max}$ (% o.r.).
- the maximum permissible measurement error (measurement accuracy) of characteristic parameter sensor, is expressed as a percentage of its full scale, respectively E_S (% o.f.s.).

On the other hand, it is remembered that supplementary, specifically only for the turndown expression of the extended reaction flowmeters:

ΔQ_m – the overlap zone of ranges of the two component reaction flowmeters ("the main", and "the secondary") of the extended reaction flowmeters

$Q_{M,max}$ – maximum value of the mass flow rate, measured by "the main" reaction flowmeter

$Q_{S,min}$ – minimum value of the mass flow rate measured by the "secondary" reaction flowmeter

Major practical utility of these expressions consists in ensuring analytical bases for obtaining optimal correlation between the necessity to maximize the turndown and a corresponding reasonable value of the reaction flowmeters accuracy, expressed by $E_{F,max}$ (% o.r.).

In conclusion, the above presentation provides the analytical bases for the evaluation of the potential achievable evolution of the performances of each basic type of reaction flowmeter, and to choose its optimal applications.

10.2 Comparative Evaluation of the Values of the Binomial (Turndown – Measurement Accuracy)

The second stage deepens the first stage (which reiterates the analytical bases of the analysis) and consists in the concrete numerical evaluation of these essential features, ensured by the comparative presentation of several numerical case examples, logically correlated, to cover entire potential ranges of their values.

These case examples include all different characteristic parameters (respectively all the different types of the sensors that measure them) used by the three basic types of the reaction flowmeters.

This synthetic overview is obtained by computing the turndown values, using the above relationships, corresponding to the scanning of the preset values ranges of the accuracy E_S of sensors that measure the characteristic parameter of the reaction flowmeters and respectively to their accuracy $E_{F,max}$.

The numerical evaluations performed correspond to the variation ranges of the measurement accuracy of the industrial sensors, usable to measure the characteristic parameters of the reaction flowmeters.

The results of these concrete numerical case examples are correlated and are systematically rendered in Table 10.1.

Based on these numerical results, an intuitive graphical overview rendering the turndown variation for each of these three basic types of the reaction flowmeters is shown in Figure 10.1, depending on the maximum permissible error $E_{F,max}$ (% o.r.) of the reaction flowmeters.

Thus, this graphical presentation is achieved for several values of the maximum permissible error E_S (% o.f.s.) of the sensor that measures the characteristic parameter of reaction flowmeter, respectively: (3×10^{-4}, 1×10^{-3}, 2×10^{-2}) % o.f.s.

TABLE 10.1

Comparative Analysis of the Binomial (Turndown T – Maximum Permissible Error of Flowmeter, E_{Fmax}), Depending on the Maximum Permissible Error E_S of the Sensors That Measure the Characteristic Parameters, for the Different Basic Types of Reaction Flowmeters

Characteristic parameter (output parameter of Primary Element and input parameter of Secondary Element) used by the Reaction Flowmeters	Type of characteristic parameter sensor of the reaction flowmeters*	Reference parameters of the extended reaction flowmeters (ERF): Reference parameters of the two nonlinear reaction flowmeters, components of ERF: Maximum permissible error of characteristic parameter sensor of the reaction flowmeters	Maximum permissible error of the reaction flowmeters	Resulting theoretical turndown of the nonlinear reaction flowmeters (implicitly of each reaction flowmeter, component of ERF)	Minimum and maximum overlap zone of the ranges of the two reaction flowmeters, component of ERF	Resulting theoretical turndown of ERF	Resulting theoretical turndown of linear reaction flowmeters	Accuracy (measurement error) range of the reaction flowmeters
		E_S	$E_{F,max}$	$T_{NRF} = T_M = T_S$	$\Delta Q_m / Q_{M,max}$	T_{ERF}	T_{LRF}	E_F
		% o.f.s.	% o.r.	-	-	-	-	% o.r.
Force	Load cell sensor	3×10^{-4}	2	81,65	0,010 0,025	3.670,09 2.192,10	6.659,00	0,0003 ... 2
			1	57,73	0,010 0,025	2.112,95 1.364,06	3.332,75	0,0003 ... 1
			0,5	40,82	0,010 0,025	1.183,27 824,68	1.666,27	0,0003 ... 0,5
			0,1	18,25	0,010 0,025	281,65 228,71	333,06	0,0003 ... 0,1
		10^{-3}	2	44,72	0,010 0,025	1.381,98 944,29	1.999,87	0,001 ... 2
			1	31,62	0,010 0,025	759,76 558,51	999,82	0,001 ... 1
			0,5	22,36	0,010 0,025	408,63 320,72	499,96	0,001 ... 0,5
			0,1	10,00	0,010 0,025	90,91 80,00	100,00	0,001 ... 0,1
		10^{-2}	2	14,14	0,010 0,025	175,22 147,77	199,93	0,01 ... 2
			1	10,00	0,010 0,025	90,91 80,00	100,00	0,01 ... 1
			0,5	7,07	0,010 0,025	49,65 42,49	49,98	0,01 ... 0,5
			0,1	3,16	0,010 0,025	9,97 9,27	9,98	0,01 ... 0,1
	Force sensor with strain gauge	$1,4 \times 10^{-2}$	2	11,95	0,010 0,025	127,61 110,00	148,80	0,014 ... 2
			1	8,45	0,010 0,025	65,85 58,97	71,40	0,014 ... 1
Differential pressure	Δp sensor	2×10^{-2}	2	10,00	0,010 0,025	90,91 80,00	100,00	0,02 ... 2
		$2,5 \times 10^{-2}$	2	8,95	0,010 0,025	73,43 65,37	80,10	0,025 ... 2
Torque	Torque sensor	10^{-1}	2	4.47	-	-	-	0,1 ... 2

Note: *This sensor constitutes the input part of the Secondary Element of the reaction flowmeter, whose output signal (proportional to "the characteristic parameter" of the flowmeter) is the output parameter Q_R of the reaction flowmeter, rendering the measured mass flow rate Q_m.

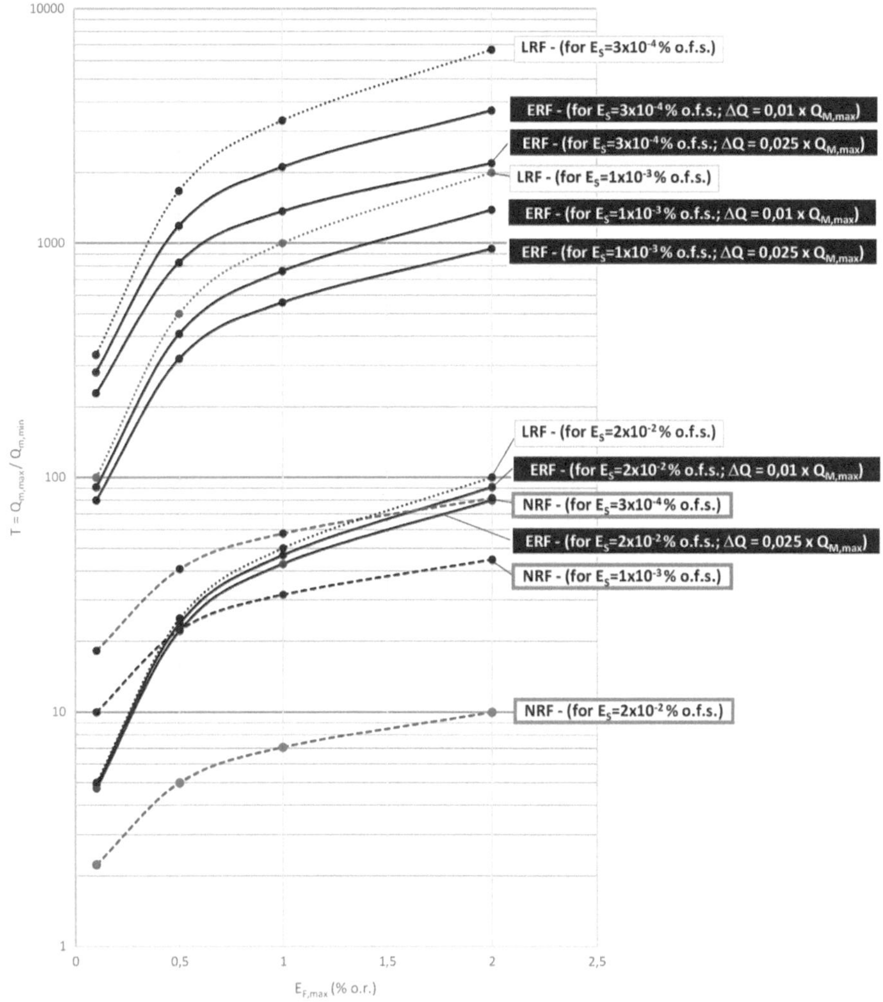

FIGURE 10.1
Evolution of the binomial (turndown T – maximum permissible error of flowmeter $E_{F,max}$) depending on the maximum permissible error E_S of the characteristic parameter sensor, for the different basic types of the reaction flowmeters.

The differentiated graphical representation of the binomial curves T ($E_{F,max}$), for each of the basic types of flowmeters, is as follows:

- dashed line represents non-linear reaction flowmeters
- dotted line represents linear reaction flowmeters
- continuous line represents extended reaction flowmeters

Also, in the same sense, every curve is provided with a tag indicating the specific values of the reference parameters of the turndown, corresponding to every numerical case example, that determine its specific appearance.

Consequently, from the theoretical analysis, presented in Table 10.1, of the values evolution of the binomial (turndown T – flowmeter accuracy $E_{F,max}$), the essential features of the reaction flowmeters can be synthetized into several main conclusions:

1. Corresponding to the same value $E_{F,max}$ = 2 (% o.r.), of the maximum permissible error of the reaction flowmeters:
   - For nonlinear reaction flowmeters (that is, reaction flowmeters without moving parts and the nonlinear dependence "mass flow rate – measured characteristic parameter"), its turndown may vary in the range (81,65... 4,47), corresponding to the decreasing of the accuracy of the characteristic parameter sensor, from $E_S = 3 \times 10^{-4}$ (% o.f.s.) to $E_S = 10^{-1}$ (% o.f.s.).
   - For linear reaction flowmeters (that is, reaction flowmeters without moving parts and the linear dependence "mass flow rate – measured characteristic parameter"), its turndown may vary in the range (6.659... 80,10), corresponding to the decreasing of the accuracy of the characteristic parameter sensor, from $E_S = 3 \times 10^{-4}$ (% o.f.s.) to $E_S = 2,5 \times 10^{-2}$ (% o.f.s.).
   - For extended reaction flowmeters (that is, a connection of two reaction flowmeters without moving parts and the nonlinear dependence "mass flow rate – measured characteristic parameter"), its turndown may vary in the range (3.670,09... 65,37), corresponding to the decreasing of the accuracy of the characteristic parameter sensors, from $E_S = 3 \times 10^{-4}$ (% o.f.s.) to $E_S = 2,5 \times 10^{-2}$ (% o.f.s.).
2. The increase of the accuracy of the reaction flowmeters (that is, decrease of the $E_{F,max}$ value) depends both the basic type of the reaction flowmeters and the sensor type of the measured characteristic parameter of the flowmeters, as follows:
   - For nonlinear reaction flowmeters, up to $E_{F,max}$ = 0,1 (% o.r.) and T_{NRF} = 18,25... 3,16 for "load cell sensors", respectively $E_{F,max}$ = 1 (% o.r.) and T_{NRF} = 8,45 for "force sensor with strain gauge".
   - For linear reaction flowmeters, up to $E_{F,max}$ = 0,1 (% o.r.) and T_{LRF} = 333,06... 9,98, respectively $E_{F,max}$ = 1 (% o.r.) and T_{LRF} = 71,40 "force sensor with strain gauge"
   - For extended reaction flowmeters, up to $E_{F,max}$ = 0,1 (% o.r) and T_{ERF} = 281,65... 9,27 for "load cell sensors", respectively $E_{F,max}$ = 1 (% o.r.) and T_{ERF} = 65,85... 58,97 for "force sensor with strain gauge".

The above synthetic comparative presentation puts in evidence the advantages offered by the reaction flowmeters to ensure, for several basic types, high values of their turndown, optimally correlated with good values of their accuracy.

This short overview ensures a global image of the configurations diversity of the reaction flowmeters, and of their gradual performance growing, due to both their specific basic configurations, and the possibility of using sensors with increased accuracy for the measurement of characteristic parameters of the reaction flowmeters, permanently following this improvement.

10.3 Other Main Features

This short overview is continued, after the above comparative evaluation of the binomial (turndown – accuracy), with a synthetic presentation of other main features of the reaction flowmeters.

A first and essential advantage of the reaction flowmeters is conceptually provided by their reaction measurement system that has a common principle configuration, universal for all basic types of them (both without and with moving parts).

Secondly, the common principle configuration of the reaction measurement system ensures:

- Simplicity of both the fluid-flowing path, for generating the reaction force, and of its takeover and measuring.
- By its flexibility, a high availability of easy takeover of the reaction force in a wide variety of ways.

Generally referring ourselves to the different basic types of the reaction flowmeters, configured so far and presented in the book, it is observed that the great majority is without moving parts (respectively, nonlinear, linear, extended, and bypass type reaction flowmeters).

This results, above all else, in all of them having the important advantage of good operational reliability, ensured by their minimal wear and tear during functioning, low maintenance, and moderate cost.

Another essential advantage of flowmeters consists in their suitability for measuring the flow rate of liquids and gases (for steam being under analysis).

Experimental calibrations with water and air of the prototypes of some basic types of reaction flowmeters have demonstrated the advantage of the possibility to establish an analytical method of conversion of the flow scale from one fluid to another.

Suspended solid particles from the fluid do not affect the measurement of the reaction flowmeters both without and with moving parts. Flowing velocities ensure the permanent self-cleaning of the reaction tube and its related connections.

Limitation of the reaction flowmeters with a quadratic dependence between the mass flow rate and the measured characteristic parameter, consists of a shorter flow rate range.

The extended reaction flowmeters, although they use the reaction flowmeters with a quadratic dependence between the mass flow and the measured characteristic parameter, achieve high values of their flow rate ranges due to their functioning algorithm.

The reaction flowmeters with moving parts, on the one hand, have good turndown values and, on the other hand, are sensitive to fluid viscosity (due to the rotation of their reaction elements), but are less sensitive than turbine flowmeters, due to the absence of the afferent blades.

The bypass type reaction flowmeters are advantageous for the measurement of the high fluid flow rates (corresponding to the conduits with large diameters), because they are cheaper and less bulky, but they have smaller accuracy than using an individual type of flowmeter.

The accuracy of the bypass type reaction flowmeters is smaller than the accuracy of the reaction flowmeter placed on the secondary conduit, that's way for this reaction flowmeter, it is useful to have a high accuracy in order to greatly attenuate the decrease of the correlated characteristics (turndown-total accuracy of the bypass type reaction flowmeters).

10.4 Conclusions

As mentioned at the beginning, this book presents all the basic types of reaction flowmeters configured by the author so far by applying his Method [2.1].

It is necessary to specify that the configuration of the reaction flowmeters presented in the book was initiated less than two years ago, so only the prototypes of different basic types of reaction flowmeters have been made and checked so far.

As this activity is in progress, further results will be added successively to the current achievements, implicitly adding to the conclusions of the above short overview on the reaction flowmeters.

Applying his Method further, the author has also analysed some new types of flowmeters (with reaction or derivatives to this principle) by obtaining connections (sometimes surprising) between the logical answers to the criteria (logical questions) provided by this Method.

In conclusion, the previous presentation referring to the configurations of these different basic types of reaction flowmeters, as a result of the first practical implementation of the Unitary Synthesis and Design Method of Flowmeters [2.1], constitutes relevant and explicit proof of the utility and efficiency of this Method and also offers the strong bases for the development of both this new method of flow measurement and the configuration of new flowmeters achieved thereof.

Selective Bibliography

1. Books

1.1. Baker, R.C. 2000. *Flow Measurement Handbook*. Cambridge: Cambridge University Press.

1.2. Cheremisinoff, C. 1987. *Flow Measurement for Engineers and Scientists*. Boca Raton: CRC Press.

1.3. Furness, R.A. 1990. *Fluid Flow Measurement*. London: Longman Book Company.

1.4. Ionescu G., V. Sgârciu, and H.M. Moțit. 1996. *Traductoare pentru automatizari industriale* (in English *"Transducers for Industrial Automation"*), vol. 2. Bucharest: Publishing House "Editura tehnica".

1.5. Miller, R.W. 1998. *Flow Measurement Engineering Handbook*. New York: McGraw-Hill.

1.6. Moțit, H.M. and A. Ciocarlea-Vasilescu. 1988. *Debitmetrie industrială* (in English *"Industrial Flow Measurement"*). Bucharest: Publishing House "Editura tehnica".

1.7. Moțit, H.M. 1997. *Contoare* (in English *"Meters – Water Meters, Gas Meters Heat Meters"*). Bucharest: Publishing House "Editura Artecno".

1.8. Moțit, H.M. 2006. *Debitmetre cu sectiune de masurare cu arie variabilă* (in English *"Variable Area Flowmeters"*). Bucharest: Publishing House "Editura AGIR".

1.9. Moțit, H.M., E. Diaconescu, and I. Făgărașan 2013. *Automatizari si instrumentie* (in English *"Automation and Instrumentation"*). Bucharest: Publishing House "Editura MATRIX ROM".

1.10. Motit, H.M. 2018. *Unitary Analysis, Synthesis and Classification of Flowmeters*. Baca Raton, FL: CRC Press/Taylor & Francis Group.

1.11. Spitzer, D.W. 2001. *Flow Measurement – Practical Guide of Measurement and Control*. North Carolina: ISA.

2. Patents

2.1. Motit, H.M. 2020. Computer-implemented Method for Unitary Synthesis and Design of Flowmeters and of Compound Gauging Structures. European Patent EN 3364159.
2.2. Moţit, H.M. 1992. Debitmetru cu inserţie (in English "Insertion Flowmeter"), Romanian Patent RO 110869.
2.3. Moţit, H.M. 1981, 1982. Metodă şi instalaţie de determinare experimentală a curbelor de conversie a scalelor de debit ale debitmetrelor cu arie variabilă (in English "Method and instalation for experimental determination of curves necessary for the analytical conversion of the scales of variable area flowmeters"), Romanian Patents RO 76751 completed by RO 79363.

3. Proceedings

3.1. Moţit, H.M. 1980. The utilization of Similitude Theory for the calculation of variable area flowmeters scales. *8th Fluidics and Fluid Engineering in Control Systems Conference, Bucharest*. Proceeding: 51–55.
3.2. Moţit, H.M. 1983. Theoretical and practical contributions to optimal design of variable-area flowmeters. *5th International Conference on Control systems and Computer Science, Bucharest*. Proceedings: 251–260.
3.3. Moţit, H.M. 1987. Metode pentru determinarea prin conversie a scarilor de debit ale debitmetrelor cu imersor articulat (in English, "Methods for determining by conversion of the flow scales of variable area flowmeters with articulated float") *3rd Romanian National Symposium of Metrology, Bucharest*. Proceeding: 70–78.
3.4. Moţit, H.M. 1994. The calibrations of the flowmeters with variable area. *The XIII IMEKO World Congress, Torino*. Proceeding: 2426–2429.
3.5. Moţit, H.M. 2016. Current status and trends in the standardization of flow measurement at international level. *22nd International Symposium of the Control and Instrumentation Association of Romania, Bucharest*. Proceeding: 12–20.
3.6. Moţit, H.M. 2019. A method of flow measurement based on the reaction force. Reaction flowmeters. *FLOMEKO 2019. 18th International Flow Measurement Conference, Lisbon*. Proceeding: 108–113.

4. Journals

4.1. Moţit, H.M. 1983. Method of optimal analysis and synthesis of the devices value, used in the design of control instrumentation. Bucharest: Publishing House "Editura Tehnică", *Automation – Management – Computers* 33: 98–136.
4.2. Moţit, H.M. 1986. The presentation of Romanian variable area flowmeters types ROTROM (Design, technical features). Bucharest: Publishing House "Editura Tehnică", *ENERG* 1: 265–301.

4.3. Moțit, H.M. 2003. Metodă de alegere computerizată a tipului specific de debitmetru cu secțiune de măsurare cu arei variabilă, care soluționează optim comanda clientului. (in English "Computer–implemented method to select the type of the variable area flowmeter which ensures the optimum solution for customer order"). Bucharest: *Automatizari si Instrumentatie* (in English, *"Automation and Instrumentation"*) 1: 11–14.
4.4. Moțit, H.M. 2006. Debitmetre neconventionale – complexe. Baze analitice și scheme structurale. (in English, "Unconventional - complex flowmeters. Analytical based and structural schemes"). Bucharest: *Automatizari si Instrumentatie* (in English *"Automation and Instrumentation"*) 5: 22–29.
4.5. Moțit, H.M. 2010. Debitmetre cu jet turbionat. Baze analitice.Caracteristici tehnice (in English, Swirl flowmeters. Analytical bases. Technical characteristics). Bucharest: *Automatizari si Instrumentatie* (in English, *"Automation and Instrumentation"*) 1: 12–17.
4.6. Moțit, H.M. 2011. Sinteza si Clasificarea Unitara a Debitmetrelor (in English "Unitary Synthesis and Classification of Flowmeters"). Bucharest: *Automatizari si Instrumentatie* (in English *"Automation and Instrumentation"*) 6: 10–21.
4.7. Moțit, H.M. 2013. Debitmetre ultrasonice pentru gaze. Necesitatea și avantajele utilizării metodei corectării exactității de măsurare cu modificarea dimensională a corpului debitmetrului (in English "Need and the advantages of using the method of correcting the metering accuracy with the dimensional changes of meter body"). Bucharest: *Automatizari si Instrumentatie* (in English *"Automation and Instrumentation"*) 1: 6–9.

5. Unpublished Documents

5.1. Moțit, H.M. 1998. The unitary structures of flowmeters. Paper presented at *ISO – TC 30 Flow measurement Meeting, Cascais – Portugal.*

6. Standards

6.1. International Standards ISO-standards developed by technical committees:
  6.1.1 TC 30 – Measurement of fluid flow in closed conduits.
  6.1.2 TC 113 – Hydrometry.
6.2. European Standards EN Standards developed by technical committees:
  6.2.1 F05 – Measuring instruments.
  6.2.2 TC 237 – Gas meters.
  6.2.3 TC 92 – Water meters.
  6.2.4 TC 294 – Communication systems for meters and remote reading of meters.
  6.2.5 TC 318 – Hydrometry.
6.3. BS 7405 (1991) – Guide to selection and application of flowmeters for the measurements of fluid flow in closed conduits.
6.4. API – Manual of Petroleum Measurement Standards: Chapter 4 Proving Systems,

Chapter 5 Liquid Metering/Metering, Chapter 6 Assemblies Metering/Metering, Chapter 14 Natural Gas Fluids Measurement.
6.5. AGA 7 – Turbine meters for gas applications.
6.6. Romanian Standards SR – Standards developed by the Technical Committee TC 119 ASRO - Measurement of fluid flow in closed conduits. Hydrometry.

7. International Metrological Recommendations (OIML Recommendations)

7.1. R-49/1, 2, 3 – Water meters for cold potable water and hot water.
7.2. R 117 – Dynamic measuring systems for liquids other than water.
7.3. R 119 – Pipe provers for testing of measuring systems for liquids other than water.
7.4. R 120 – Standard capacity measures for testing measuring systems for liquids other than water.
7.5. R 140 – Measuring systems for gaseous fuel.

Index

Note: *Italicized* page numbers refer to figures, **bold** page numbers refer to tables